PERGAMON INTERNATIONAL LIBRARY
of Science, Technology, Engineering and Social Studies

*The 1000-volume original paperback library in aid of education,
industrial training and the enjoyment of leisure*

Publisher: Robert Maxwell, M.C.

*Judy
Shroud.*

A TEXTBOOK OF
HUMAN BIOLOGY
SECOND EDITION

THE PERGAMON TEXTBOOK
INSPECTION COPY SERVICE

An inspection copy of any book published in the Pergamon International Library
will gladly be sent to academic staff without obligation for their consideration for
course adoption or recommendation. Copies may be retained for a period of 60 days
from receipt and returned if not suitable. When a particular title is adopted or
recommended for adoption for class use and the recommendation results in a sale
of 12 or more copies, the inspection copy may be retained with our compliments.
The Publishers will be pleased to receive suggestions for revised editions and new
titles to be published in this important International Library.

Other Titles of Interest

BACQ, Z. M. and ALEXANDER, P.
Fundamentals of Radiobiology

GREWER, E.
Everyday Health

PARSONS, T. R. and TAKAHASHI, M.
Biological Oceanographic Processes

WHITE, D. C. S. and THORSON, J.
The Kinetics of Muscle Contraction

WISCHNITZER, S.
Introduction to Electron Microscopy

BACQ, Z. M.
Fundamentals of Biochemical
Pharmacology

CREW, F. A. E.
The Foundations of Genetics

PARKE, D. V.
The Biochemistry of Foreign Compounds,
2nd ed.

ROLLS, E. T.
The Brain and Reward

BERENBLUM, I.
Cancer Research

LAMB, M. W. and HARDEN, M. L.
The Meaning of Human Nutrition

CEMBER, H.
Introduction to Health Physics

KOREN, H.
Environmental Health and Safety

BIRCH, G. G. *et al.*
Food Sciences

DUCKWORTH, R. B.
Fruit and Vegetables

KENT, N. L.
Technology of Cereals, 2nd ed.

LAWRIE, R. A.
Meat Science, 2nd ed.

RHODES, A. and FLETCHER, D. L.
Principles of Industrial Microbiology

THREADGOLD, L. T.
The Ultrastructure of the Animal Cell,
2nd ed.

FAHN, A.
Plant Anatomy, 2nd ed.

GOODWIN, T. W. and MERCER, E. I.
Introduction to Plant Biochemistry

GOSS, J. A.
Physiology of Plants and their Cells

LESHEM, Y.
The Molecular and Hormonal Basis of
Plant Growth Regulation

STREET, H. E. and COCKBURN, W.
Plant Metabolism

WAREING, P. F. and PHILLIPS, I. D. J.
The Control of Growth and Differentiation
in Plants

PINNIGER, R. S.
Jones' Animal Nursing. Fully revised 2nd
ed.

The terms of our inspection copy service
apply to all the above books. Full details
of all books listed will gladly be sent upon
request.

J K INGLIS
BSC, BA, DIP ED, MI BIOL

*Lecturer in Biology, College
of Further Education, Oxford*

*Sometime Instructor
Anatomy and Physiology,
College of Lake County, Illinois, U.S.A.*

A Textbook of
HUMAN
BIOLOGY
SECOND EDITION

PERGAMON PRESS

*Oxford · New York · Toronto
Sydney · Paris · Frankfurt*

U.K.	Pergamon Press Ltd., Headington Hill Hall, Oxford OX3 0BW, England
U.S.A.	Pergamon Press Inc., Maxwell House, Fairview Park, Elmsford, New York 10523, U.S.A.
CANADA	Pergamon of Canada, Suite 104, 150 Consumers Road, Willowdale, Ontario M2 J1P9, Canada
AUSTRALIA	Pergamon Press (Aust.) Pty. Ltd., P.O. Box 544, Potts Point, N.S.W. 2011, Australia
FRANCE	Pergamon Press SARL, 24 rue des Ecoles, 75240 Paris, Cedex 05, France
FEDERAL REPUBLIC OF GERMANY	Pergamon Press GmbH, 6242 Kronberg-Taunus, Pferdstrasse 1, Federal Republic of Germany

First edition 1968

Reprinted 1969

Reprinted (with corrections) 1971

Reprinted 1973

Second edition 1974

Reprinted 1976 (twice), 1977

Reprinted (with corrections) 1979

Library of Congress Catalog Card No. 73-21574

Printed in Great Britain by A. Wheaton & Co. Ltd, Exeter

ISBN No. 0 08 017846 4 (hard)

ISBN No. 0 08 017847 2 (flexi)

Contents

x   *Contents*

Preface to the Second Edition

THIS second edition is a much revised and rewritten version of the first edition. It is intended for the C.S.E. and the G.C.E. student as well as those studying the basic courses of Technician Education Council. (previously O.N.C. and City and Guilds of London).

Many students studying Human Biology move on to nursing, medical laboratory work or university courses in medicine. To this end some disorders and diseases have been included within each chapter. To study the normal body function without some mention of its abnormalities would be somewhat short-sighted and fail to give a perspective to the "wonders" of the normal body.

An added dimension to the study of the Human animal has been included in this edition, that of practical work. Laboratory work in Human Biology is frequently neglected and so a few simple exercises have been suggested. The "hygiene" section of the book has been kept as brief as possible but this does not imply any inferior status. Hygiene study without human reference can be rather dull and so project work in this field is strongly recommended.

Introduction to the Species

MAN is a mammal because he has hair and mammary glands. The human species walks upright on two feet, is land dwelling, very sociable and omnivorous in diet.

Man tends to be more advanced than other animals in some or all of the following:

(a) He has a large and complex brain.
(b) He has an unspecialized hand with opposing fingers and thumb useful for accurate use of tools.
(c) He makes and uses tools.
(d) He has stereoscopic vision.
(e) He uses language and speech for communication.
(f) He modifies his environment for his own purposes.
(g) He creates and survives in social institutions.
(h) He forms "mental" and abstract concepts.
(i) He often cooks his food.

Modern man, *Homo sapiens,* is probably 100,000 years old. His origins are very difficult to disentangle, but it is thought that he emerged from man-like apes living during the Miocene and Pliocene geological periods (Fig. 1). At the moment we have found no human remains dating before the Pleistocene period.

During the time when man was making his appearance the northern parts of Europe and America were covered in a great, slow-moving glacier. This massive frozen belt around the world naturally influenced the climate of that time and therefore the plant and animal life which was able to survive. Near the ice it was cold, as the ice moved away

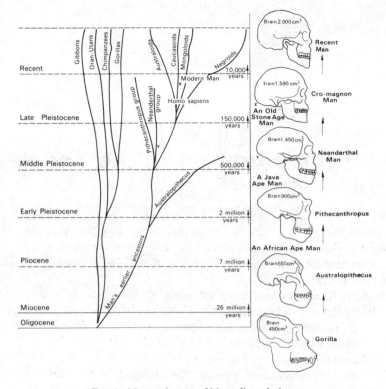

FIG. 1. Man and some of his earlier relatives.

it was warmer. The warmer areas encouraged plant growth and so also animal life. No plants—no animals. Our Pleistocene ancestors may have looked like the artist's impression in Fig. 2. Remember we only have bony relics upon which to build this picture. It could be exaggerated.

Early man shared the continents with many of the plants and animals that live today, together with many that are now extinct—the great mammoths, rhinoceros, bears and lions—even in Europe. His animal companions, the rock tools that he made, the plants and other things buried with or near his skeleton help us to date and understand a little of early man.

Homo sapiens is a species, all races of which are capable of inter-

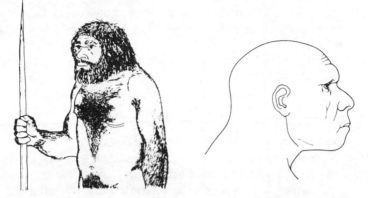

FIG. 2. An "old stone age" man—Neanderthal man.

breeding. The races of modern man are sometimes classified upon the basis of hair type and his skull measurements. One classification of modern man is as follows:

1. *Negriforms*. Long, narrow-headed (dolichocephalic). Woolly, spiral hair. Flattened noses. Skin usually dark.

   This group includes the negroes of Africa and other negro types in Malaysia and other areas.

2. *Europiforms*

   (a) *Mediterranean types*. Long-headed with wavy hair. Straight narrow noses. Dark eyes.

   This group includes the north African Arabs and many south Asians.

   (b) *Alpine types*. Broad-headed (brachycephalic) with darkish wavy hair. Thick-set people.

   This group includes the Slavs of eastern Europe and large groups within Russia.

   (c) *Nordic types*. An intermediate shaped head with a tendency towards narrowness. Reddish-white skin, wavy, light-coloured hair. A straight nose with a pronounced chin and blue eyes.

   This group includes the Scandinavians and many in northern Europe.

3. *Mongoliforms.* Broad-headed types with straight dark hair. They have yellowish or reddish skin. They have intermediate width noses.

   This group includes those people in the greater part of eastern Asia and the Eskimos of the north. Some of these types have the "slit" eye with the epicanthic fold on the upper lid.

4. *Australoforms.* Long-headed with hair. They have broad noses like those of the negroes. They have dark complexions. They are the original inhabitants of Australia.

Man as a mammal must relate much of his success as a land dwelling animal to his reproductive methods. Man, like other mammals, deposits sperm internally into the female (internal fertilization). This sperm is thereby able to fertilize the egg without the hazards that accompany external fertilization which is practised by such as fish. This practice allows the human female to reduce the number of eggs produced to one or two per month instead of the millions produced by the fish.

Internal development of the young similarly is a protective measure that allows a more advanced creature to be born. There follows a long period of "mothering". It is this "mothering", its length and quality which sets man apart from other mammals.

CHAPTER 1

Cells, Tissues and Organs

CELLS

Man's body is made of *protoplasm*. Protoplasm is a "blanket term" covering that substance which shows characteristics which we call life. Protoplasm may be semi-solid or in various different forms depending upon from which part of the body we take it. It is within the protoplasm (or its derivatives) that all the chemical reactions of life take place.

Protoplasm is organized into units called *cells*.

Cells are composed of the following structures:

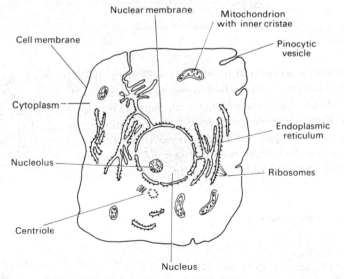

Fig. 3. Animal cell structures (diagrammatic), showing main organelles.

1

Nucleus, made up of a jelly-like *nucleoplasm* in which are suspended thread-like *chromosomes* (becoming obvious at cell division) and one or more *nucleoli.* The whole is enveloped in a *nuclear membrane.*

(a) The *chromosomes* are made up largely of protein and nucleic acids—*nucleoproteins.* The nucleic acid part of the chromosome is DNA (deoxyribonucleic acid). Chromosomes are the carriers of genes which regulate all our cellular processes and are responsible for our inherited characteristics. The numbers of chromosomes in the nucleus are typical for the species. Human nuclei have forty-six chromosomes arranged in twenty-three pairs. These chromosomes are diffused within the nuclear sap as a granular looking *chromatin* which stains blue with a dye such as haematoxylin. The chromosomes come into view during cell reproduction, called *mitosis* or *meiosis.*

(b) *Nucleoli* are also made up of nucleoprotein but in this case the nucleic acid is RNA (ribonucleic acid).
They are concerned with protein manufacture within the cell.

(c) *The nuclear membrane* is a double structure made up of fat (lipid) and protein. It is perforated by minute pores to allow two-way transport.

Cytoplasm, made up of a granular substance in which are suspended many *organelles.* The organelles vary depending upon the type of cell and its function.
Some commoner organelles are:

(a) *Mitochondria* (a mitochondrion) are scattered throughout the cytoplasm of all animal cells. They may be thread-like or rod-shaped structures with a double membrane, the inside membrane being infolded to produce finger-like processes called *cristae.*
The mitochondria are the centres of cellular respiration, that is they produce the energy needed by the cell.

(b) *Ribosomes* are tiny nucleoprotein granules containing the RNA. They are located along some outer edges of the maze of membranes running throughout the cytoplasm. The ribosomes contain enzymes which are needed in the manufacture of proteins. It is at the ribosome that protein building takes place.

(c) *Centrioles* lie just outside the nuclear membrane and are involved in the process of cell division as will be seen later. They link up with the "fibres" which are related to the movement of chromosomes when the cell divides in half.

(d) *Endoplasmic reticulum* is a network of double membranes linked up with the nuclear membrane and the outer plasma, or cell membrane. This "cell-skeleton" of membranes supports the cytoplasmic "jelly". Ribosomes and other cell organelles are linked to this membraneous maze running throughout the cytoplasm.

(e) *Golgi-bodies* are not unlike the membraneous structures just mentioned but they are not seen in all cells and may even be invisible depending upon the method used to prepare the microscope slide. Golgi bodies tend to be sac-like and are found mostly in cells which are producing secretions, their sac-like membranes containing the material to be secreted.

Cell membrane (plasma membrane), a double lipid-protein membrane continuous with the nuclear membrane and the endoplasmic reticulum. This continuity can be seen in one place on the cell in the diagram (Fig. 3). Notice that this membrane has no outer strengthening material as one finds in the case of plant cells (no cellulose). The plasma membrane must not be simply thought of as a sort of plastic covering like that surrounding foods in the deep-freeze. The plasma-membrane is a living structure which needs food and oxygen like any other living structure. It has some very surprising and complex jobs to do, many of which are not clearly understood at the time of writing. The biggest mystery surrounding the cell membrane is how does it regulate the traffic of chemical materials back and forth through its tiny pores? It takes in some things but rejects others. It is known how some bulk materials get into a cell; pinocytosis and phagocytosis.

Pinocytosis (cell-drinking). A fine tunnel forms a pinocytic vesicle; the fluid in it becomes trapped within the cell as the cell membrane comes together behind it.

Phagocytosis (cell-eating) can be seen in animals like *Amoeba* where the cytoplasm flows around a small particle which becomes trapped within the cell. White blood cells, it will be seen, destroy invading bacteria by "eating" them in this way.

Animal and Plant Cells Compared

The cells of animals and plants are similar in many respects differing essentially in the items mentioned in Table 1.

TABLE 1

Animal cells	Plant cells
(a) A plasma membrane without a cellulose covering.	(a) A plasma membrane with an outer non-living cellulose covering.
(b) Chloroplasts not found within the cytoplasm.	(b) Chloroplasts, chlorophyll-containing structures, found within the cytoplasm. Chlorophyll is involved in the plant's sugar-producing process called photosynthesis.

Animals and plants differ in the way in which they obtain food. Animals, it will be seen, are dependent upon plants for their existence. The reasons for this dependence are as follows:

(a) *Animals need carbohydrates* (starches and sugar) but only plants can produce them. Plants can manufacture carbohydrate in the presence of sunlight by a process described as *photosynthesis*. Plant cells containing chlorophyll are only able to do this.

(b) *Animals need nitrogen compounds* (proteins, etc.) but are unable to produce them from simple sources of nitrogen such as the atmosphere or water soluble nitrogen salts. Plant cells (with the assistance of bacteria in some cases) are able to manufacture nitrogen compounds which are needed by animals.

The carbon cycle and the nitrogen cycle (Fig. 4) are ways of drawing out this dependence of animals upon plants. They both show how closely the livelihoods of animals and plants are linked.

The Characteristics Exhibited by Living Cells

Living cells make up living organisms such as man. The characteristics of these living cells are also the characteristics of living things in general. The seven main characteristics are shown in Table 2.

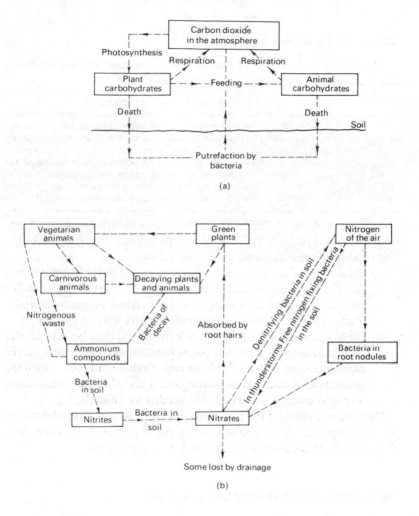

FIG. 4. (a) The carbon cycle. (b) The nitrogen cycle.

TABLE 2

	Cells	Man
Movement	Whole cells may move. The cytoplasm "flows about" the plasma-membrane.	Skeletal muscles move bones which pivot around joints. "Soft-parts" (viscera) such as heart and intestines are constantly moving.
Sensitivity	Cells can change in nature, move or react in some way to stimuli such as pressure, chemicals or radiations.	Most tissues of man's body have sense-organs to give information about conditions outside or within the body. Sense organs are sensitive to light (eye), vibrations in air (ear), temperature changes, etc.
Respiration	Cells need energy and this energy is liberated by chemical reactions which need oxygen. Cellular respiration takes place in the mitochondria.	Man needs energy and this is provided by respiration within the cells. Taking air into the lungs is referred to as external respiration because it supplies the oxygen needed for the internal or cellular respiration.
Nutrition	Cells need nutrients such as sugars for energy, proteins for building. Cells take up these nutrients across their plasma membranes.	Man's needs are those of his cells. Eating and drinking provides energy and building materials, amongst other things, for his cells. The organs associated with nutrition are specially designed to deal with our *omnivorous* diet.
Growth	Cells usually grow to an optimum size. This size is usually related to the ratio of cell membrane area to bulk within the membrane. Problems are created if the nucleus is separated by great distances from the outer cell membrane. Growth to an optimum cell size followed by mitosis (cell division) happens in many parts of the body.	Growth in man is regulated by hereditary factors (genes) carried on the chromosomes and secondarily by chemical messengers (hormones) circulating within the blood. Tissues grow by cell multiplication called mitosis. Unnatural mitotic activity in tissues produce tumours or cancers. The causes of this unnatural behaviour, or mitosis in tissue where it is inappropriate is the subject of much cancer research.

TABLE 2 *(Cont.)*

	Cells	Man
Excretion	Some by-products of chemical activity within cells must be eliminated otherwise their accumulation may poison the cell. Common excretory products are: Carbon dioxide Water Urea	The body has specialized areas for the elimination of excretory products produced as a result of cellular chemistry (metabolism). The main areas of excretion are: The Kidneys The Skin The Lungs
Reproduction	Cells can increase their numbers, that is reproduce, in two ways. *Mitosis*—cell division to produce identical "offspring". This growth takes place in areas such as at the base of the layers of the skin epidermis. *Meiosis*—"reduction" cell division is a special sort of cell reproduction which only takes place in sex organs to produce sex cells such as sperms or eggs.	Man reproduces his kind by uniting two cells, one the sperm from the male and the other the egg from the female. Sexual reproduction involves the two sexes uniting and sperms from the male are put into the female. Fertilization may result and a baby begins to grow.

The subject of Human Biology is to expand upon the topics just outlined in Table 2. Human Biology in this context is therefore

Anatomy — study of body structures,

Physiology — study of body functions.

The Way in which Cells Reproduce—Mitosis and Meiosis

Firstly, what is the difference between mitosis and meiosis? In order to understand this properly we must think back to those thread-like structures called chromosomes which are in every cell nucleus.

Each species of animal, it was said, has a definite number of these chromosomes, for example man has twenty-three pairs.

The main differences between mitosis and meiosis are illustrated in Fig. 5.

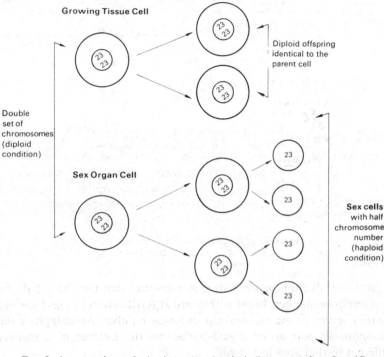

FIG. 5. A comparison of mitosis (top) and meiosis (bottom) (where $2n=46$) (n=chromosome number).

MITOSIS

Cell division of this type is similar to meiosis in that it can be organized, for our convenience, into four phases—prophase, metaphase, anaphase and telophase.

The main point of difference to be noted is the sequence of events happening at metaphase. (see Figs. 5 and 6).

In mitosis at metaphase (when chromosomes line up across the middle of the nucleus) the chromosomes are split into chromatids and the centromeres are all in the same plane. The result of mitotic division is two cells which are the same in genetic character as the parent.

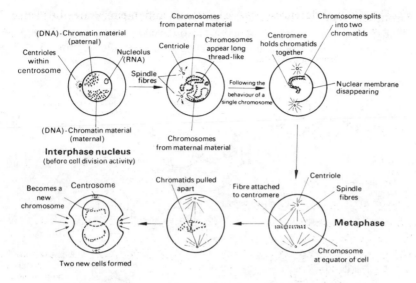

Fig. 6. Some stages in mitosis (one chromosome only shown).

MEIOSIS

In the production of sex cells it is important that they have half the chromosome number, because they are at fertilization to combine and thereby return to the parental chromosome number. At metaphase the *chromosomes pair up on a mid-line* across the nucleus in a manner different to that seen in mitosis.

The important point of difference between mitosis and meiosis is the halving of chromosome numbers and the redistribution of genetic material (genes) in order to produce variation in the character of the sex cells and eventually the differences between people.

Cellular Metabolism

Metabolism is the sum total of chemical activity within the living organism, involving respiration, nutrition and synthesis of new materials. Metabolism involves build-up (anabolism) and break-down (catabolism) of chemical substances. All these processes are regulated in order to maintain a *steady state* (*homeostasis*) and any deviation from "normal" in the human body is immediately corrected to bring things back to normal—for example our response to overheating is sweating, so as to cool the body down to comfortable temperatures.

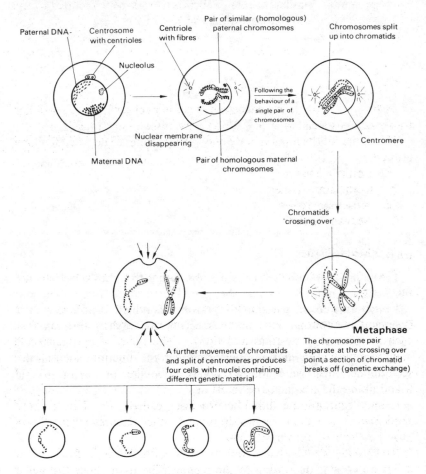

FIG. 7. Some stages in meiosis (one chromosome pair only shown).

Apart from keeping things "normal" the body must also replace dead tissue and it must reproduce the species. Metabolic processes within the cell synthesize the units of heredity, that is the nucleic acids such as deoxyribonucleic (DNA) (genes).

Basal metabolic rate (BMR) is the rate of chemical activity at rest. Each person has a different rate of chemical "tick-over" related to the state of health, glandular condition and so forth.

TISSUES

Definition. A tissue is a collection of similar cells which are modified to perform a special task, e.g. blood cells, bone tissue, muscle.

The tissues of the human body may for convenience be placed in *four* groups:

   (i) Epithelial tissues.
  (ii) Connective tissues.
 (iii) Muscular tissues.
 (iv) Nervous tissues.

(i) Epithelial Tissues

These tissues cover surfaces in contact with space, inside and outside the body.

Simple epithelium is a single layer of cells which is attached to a basement membrane. This tissue lines internal organs and cavities, such as stomach, intestines and thorax. The types of epithelial cell may be modified in shape to perform a certain function, such as the secretion of digestive fluids. It may be cube-like, or column-like if found lining the stomach and intestine.

It may bear minute hair-like processes called *cilia*. This *ciliated epithelium* is found in the windpipe and other cavities where fluids must be "lashed" into movement.

Stratified epithelium has many layers of epithelial cells, the number of layers varying according to the region. This tissue lines the outer surfaces of the body, the mouth and the gullet. The top cells are usually scaly and dead, so that some friction can take place without damage to living cells. Much of the heel skin consists of dead cells of stratified epithelium.

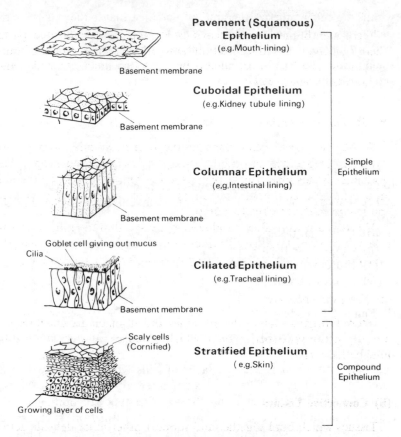

FIG. 8. Types of epithelia.

MEMBRANES

Organs and cavities may be enveloped or lined by tissues called membranes. The function of these membranes is to protect, to keep moist or to absorb substances. They can also produce a lubricating substance to reduce friction between moving parts. There are three important membranes.

(a) *Synovial membranes*

These membranes line the cavities of joints and associated parts. Their function is to reduce friction between the ends of moving bones or tendons. The friction is reduced by a thickish fluid rather like egg-white called *synovial fluid*.

(b) *Mucous membranes*

This moist type of membrane lines the alimentary canal from mouth to anus, the bladder and urethra and the respiratory passages. The epithelial tissues lining these regions vary from pavement epithelium in the mouth, columnar epithelium in the intestines and ciliated epithelium in the trachea (windpipe).

Mucus is a viscous fluid produced by cells within the mucous membrane, it protects the digestive tract from damage, it also moistens the cavity of the guts and related organs such as stomach and bladder.

(c) *Serous membranes*

Internal organs such as the intestines and stomach are enveloped in a serous membrane (peritoneum) which produces a watery lubricating fluid called *serum*.

(ii) **Connective Tissues**

Tissues which bind together and support other more delicate active parts are connective tissues. The structure of connective tissue varies from the fluid blood (dealt with separately under Chapter 6) to the strong, inflexible bone. Some of the more important connective tissues are mentioned below.

Fibrous tissue. There are two main types of binding fibre:

White non-elastic fibres are found in tendons and ligaments. These fibres run in wavy bundles throughout many connective tissues. They do not stretch.

Yellow elastic fibres are found in the walls of arteries, respiratory tubes and lungs. Many organs which are liable to change size or shape contain these elastic fibres.

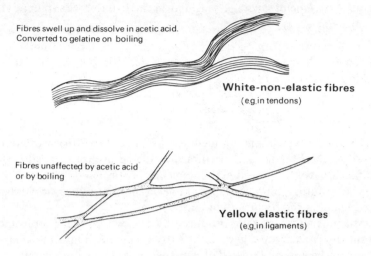

Fibres swell up and dissolve in acetic acid. Converted to gelatine on boiling

White-non-elastic fibres
(e.g. in tendons)

Fibres unaffected by acetic acid or by boiling

Yellow elastic fibres
(e.g. in ligaments)

FIG. 9. Connective tissue fibres.

Areolar tissue. This type of tissue is a loose, supporting and packing tissue found under the skin and mucous membranes. White and yellow fibres run throughout.

Adipose tissue. This is fatty tissue resembling areolar tissue with many cells swollen with storage fat. Fatty tissue acts as a heat insulator and as a reserve of energy food. It is found around the intestines, kidneys, heart and under the skin.

Cartilage (gristle). A bluish-white flexible supporting tissue. Found mainly in association with the skeleton. In most places it changes to bone as a child grows up.

Cartilage can be of three types:

Hyaline cartilage is a clearish background (matrix) produced by cartilage cells which are commonly in pairs. It is found as a flexible, protective covering for the ends of bones at joints. It is also found in the ribs, windpipe and voice box.

Fibro-cartilage is hyaline cartilage with white non-elastic fibres running through it. This type of cartilage makes up the intervertebral discs.

Elastic cartilage is cartilage with yellow elastic fibres running through it. This type of cartilage is found in the external ear (pinna) and the epiglottis.

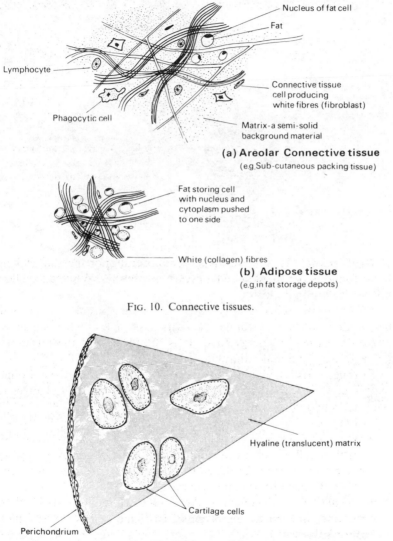

Nucleus of fat cell

Fat

Lymphocyte

Connective tissue cell producing white fibres (fibroblast)

Phagocytic cell

Matrix-a semi-solid background material

(a) Areolar Connective tissue
(e.g.Sub-cutaneous packing tissue)

Fat storing cell with nucleus and cytoplasm pushed to one side

White (collagen) fibres

(b) Adipose tissue
(e.g.in fat storage depots)

FIG. 10. Connective tissues.

Hyaline (translucent) matrix

Cartilage cells

Perichondrium

FIG. 11. Hyaline cartilage.

Bone is the hardest of connective tissues. Mineral salts such as calcium phosphate and calcium carbonate are mainly responsible for this hardness.

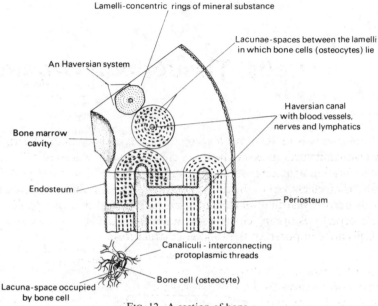

Fig. 12. A section of bone.

Microscopic examination of bone will show it to be made up of minute concentric units called *Haversian* systems each penetrated by a canal. These systems and canals run lengthwise along the bone. The rings of bony substance are termed *lamelli* with spaces called *lacunae* containing the bone cells (osteocytes). Protoplasmic connections running from one lacuna to another are called *canaliculi*. The substance of bone has a blood supply within the Haversian canals.

(iii) Muscular Tissues

There are three main types of muscular tissue:
Striped voluntary muscle has elongated striated cells. These muscles are attached to the skeleton and are under our conscious control.

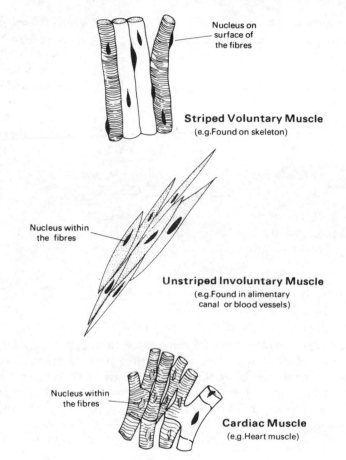

Nucleus on surface of the fibres

Striped Voluntary Muscle
(e.g.Found on skeleton)

Nucleus within the fibres

Unstriped Involuntary Muscle
(e.g.Found in alimentary canal or blood vessels)

Nucleus within the fibres

Cardiac Muscle
(e.g.Heart muscle)

FIG. 13. Types of muscle.

Unstriped (smooth) involuntary muscle has unstriated cells by which internal (visceral) organs are moved. These muscles are not under our conscious control.

Cardiac muscle is a striated and much branched type of muscle found only in the heart. This muscle is active throughout life and has a rather special nervous control system.

(iv) Nervous Tissue

The unit of the nervous system is called a *neurone*.

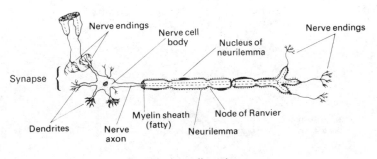

FIG. 14. A myelinated nerve.

The neurone is made up of a cell body with branch-like *dendrites*. It is these dendrites which *synapse* with the nerve endings of other nerves. From the cell body there is an extended *axon* or nerve fibre which terminates in a much branched nerve ending. The axon is usually insulated by a fatty sheath called a *myelin sheath*. The sheath is enveloped within a *neurilemma*.

ORGANS

Groups of tissues performing special functions are called *organs*. The structure of the organ is related to the type of function required of it. The eye is photosensitive, the ear sensitive to air vibrations.

There are organs related to all the major functions of the human body already mentioned. More detailed study of the organs and their structure will be deferred until our study of the appropriate organ system, for example excretory system, reproductive system, etc.

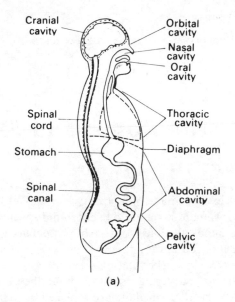

Cranial cavity

Orbital cavity

Nasal cavity

Oral cavity

Spinal cord

Thoracic cavity

Stomach

Diaphragm

Spinal canal

Abdominal cavity

Pelvic cavity

(a)

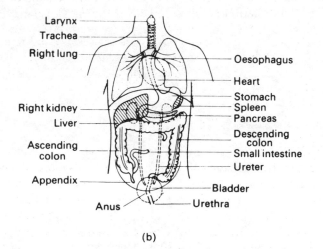

Larynx

Trachea

Right lung

Oesophagus

Heart

Stomach

Right kidney

Spleen

Liver

Pancreas

Descending colon

Ascending colon

Small intestine

Ureter

Appendix

Bladder

Anus

Urethra

(b)

FIG. 15. (a) Body cavities. (b) Organs and their location.

CHAPTER 2

The Bones and Skeleton

BONES

The adult skeleton is made of a strong tissue called bone. Bone has cells scattered throughout the rigid background substance of which it is made. This ground substance consists of protein fibres with insoluble mineral salts deposited among them. These salts are usually calcium phosphate and calcium carbonate.

Bone appears in the young child at about the second month of pregnancy when areas of cartilage are replaced as *ossification* (bone-building) begins. This process is not finished when the child is born because hand and feet small "bones" are still cartilage in structure. Similarly, ossification is not complete at the ends of bone, called the *epiphyses*.

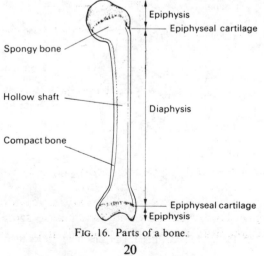

Epiphysis
Epiphyseal cartilage
Spongy bone
Hollow shaft
Diaphysis
Compact bone
Epiphyseal cartilage
Epiphysis

FIG. 16. Parts of a bone.

20

It is at these epiphyseal cartilages that bones grow to their adult length. They become ossified at the cessation of long bone growth between the years of 18 and 25.

The Structure of Bone

The microscopic structure of compact bone has been dealt with previously (Fig. 12).

The general structure of a long bone shown in section will reveal the following main parts:

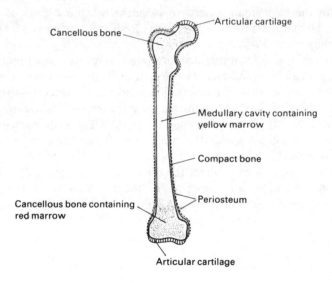

Cancellous bone

Articular cartilage

Medullary cavity containing yellow marrow

Compact bone

Cancellous bone containing red marrow

Periosteum

Articular cartilage

FIG. 17. A long bone.

(a) The *periosteum* is the outer fibrous tissue which covers the whole of the bone, except the articular cartilage. Blood vessels and nerves run in this membrane and from here they pass to the outermost layers of bone only. Bone has no sense organs, the **periosteum picks up sensations from this area.**

(b) The *compact bone* lies beneath the periosteum. It is the outer layer of dense hard bone.
(c) The *cancellous* (spongy) bone is a porous tissue which is replaced by blood forming tissue, red marrow at the ends of the bone.
(d) The *bone marrow* is found within the medullary cavity as a yellow fatty tissue during adult life. Red bone marrow tends to occupy the ends of bones.

The Growth of Bone

Bones grow by the deposition of insoluble calcium phosphate beneath the periosteum (increase in girth) or at the ends of bones where the diaphysis joins with the epiphysis (increasing the length of the bone).

This 'depositing' of mineral salt at points of growth is a function of bone cells called *osteoblasts*. They produce an enzyme which converts blood-carried soluble phosphates into insoluble calcium phosphate. Within the cavity of the bone the reverse process is taking place thereby breaking down older bone material. The cells carrying out this task are called *osteoclasts*. Over a given time a bone will renew its mineral content.

The growth of bone is under the direction of hereditary factors and hormones. Vitamin D is important in the diet because of its indirect influence upon the maintenance of the correct calcium level in the blood.

Types of Bone

Bones may be classified according to shape.

Long bones are usually cylindrical with a knob at each end. They usually have a hollow shaft, e.g. femur (vertebrae are compressed long bones).

Short bones are spongy bone units covered by a thin layer of compact bone, e.g. carpal and tarsal bones of hands and feet.

Flat bones are plates of compact bone with a sandwich of spongy bone between, e.g. sternum, skull bones (many), parts of the pelvis.

THE SKELETON

The bones of the skeleton can be grouped as follows:
(a) *Axial skeleton*
    (i) Skull: cranium, facial bones.
    (ii) Backbone and rib cage.
(b) *Appendicular skeleton*
    (i) Limb girdles: pectoral and pelvic.
    (ii) Limbs: arms and legs.

The human skeleton is made up of 206 bony units bound together by strong fibrous *ligaments*.

In discussing the skeleton no detailed information is given about individual bones. This detail is best obtained by handling and examining the individual bones in the laboratory.

The study of the skeleton will commence from the skull moving downwards to the lower limbs.

(i) The Skull

The bones of the skull are mostly flat and come close together at joints called *sutures*; they are generally immovable.

The skull is mounted upon the atlas vertebra which allows movement backward and forward.

The cranial bones form a cavity which contains and protects the brain. The eight bones which make up the cranium are not fused at birth. This allows for the head pressures experienced during birth and for subsequent growth. Two areas on the *cranium* of the new-born child are soft and may be seen to pulsate. These areas are the *fontanelles*.

The facial bones, fourteen in number, are attached to the cranium. They support the muscles of the face, mouth and nose.

The upper jaw consists of two *maxillae* which meet at a mid-line below the nose.

The lower jaw or *mandible* is a movable structure articulating with the temporal bone in front of the ear. The maxillae and the mandible both carry teeth.

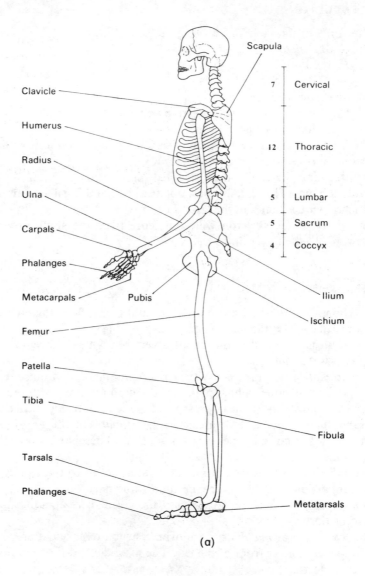

FIG. 18. The skeleton. (a) Side view.

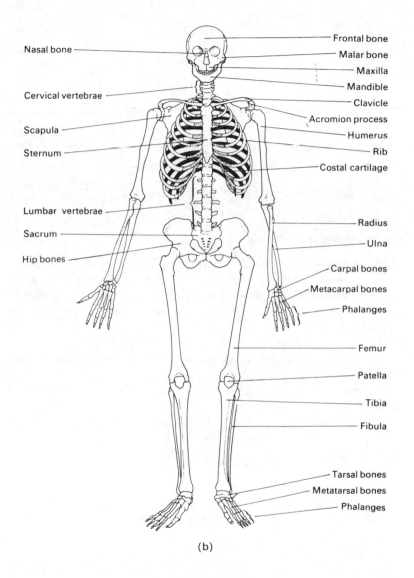

Nasal bone

Cervical vertebrae

Scapula

Sternum

Lumbar vertebrae

Sacrum

Hip bones

Frontal bone

Malar bone

Maxilla

Mandible

Clavicle

Acromion process

Humerus

Rib

Costal cartilage

Radius

Ulna

Carpal bones

Metacarpal bones

Phalanges

Femur

Patella

Tibia

Fibula

Tarsal bones

Metatarsal bones

Phalanges

(b)

FIG. 18. (b) Front view.

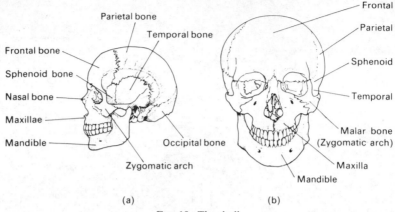

Fig. 19. The skull.

(ii) The Pectoral Girdle and Upper Limbs

The *pectoral girdle* consists of a pair of *clavicles* and a pair of *scapulas*.

The clavicles. These are a pair of collar bones forming the front part of the shoulder girdle (pectoral girdle). One end of each curved bone articulates with the sternum. The other end articulates with the scapula.

The scapulas. These are the two shoulder blades, triangular bones forming the back part of the pectoral girdle. They do not articulate with the ribs but are embedded in muscle.

A depression called the *glenoid cavity* accommodates the ball end of the upper arm bone.

The upper limb consists of the following bones:

The humerus: upper arm bone.

The radius ⎱ forearm bones.
and ulna ⎰

The carpals (carpus): eight short bones forming the wrist.

The metacarpals (metacarpus): five short bones forming the palm.

The phalanges: three bones forming the fingers. Two bones form the thumb.

The humerus. The upper end of this long bone has a hemi-spherical-shaped *head* which makes a rather shallow ball and socket joint with the glenoid cavity of the scapula.

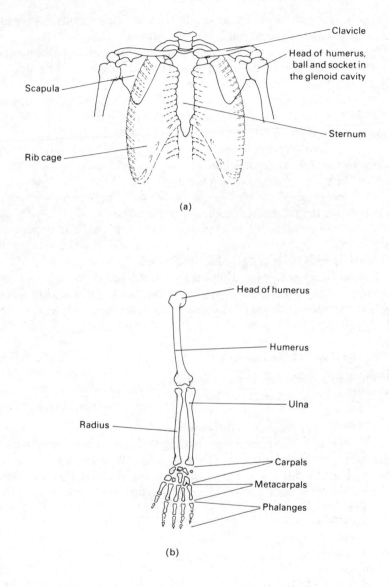

(a)

(b)

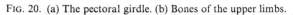

FIG. 20. (a) The pectoral girdle. (b) Bones of the upper limbs.

The lower end of the humerus has projections (condyles) which articulate with the radius and ulna of the forearm.

The shaft of the bone has a rough surface for muscle attachments.

The radius. This is the outer bone of the forearm. The upper, smaller end articulates with the humerus.

The lower end of the bone is wider and articulates with the carpus.

The ulna. This is the inner and larger bone of the forearm. At the upper end of the bone it forms a hinge joint with the humerus.

The *olecranon process* prevents the elbow joint from bending backward and acts as an attachment for the triceps muscle. This process causes the point of the elbow on the upper end of the ulna.

The lower end of the ulna is smaller than the upper end, and articulates with the radius. It does not join the carpus.

The carpus consists of two rows of four short bones. The upper row articulates with the radius and forms the wrist joint. The lower row articulates with the metacarpals of the palm.

The metacarpus consists of five small bones making up the palm of the hand. The lower ends of the metacarpal bones form the knuckles, which articulate with *the phalanges* or finger bones.

(iii) The Bones of the Trunk

The bones of the trunk are as follows:
The sternum: the breast-bone.
The ribs.
The vertebral column: the backbone.

The sternum. The breast-bone is shaped like a dagger, from which the names of the two main parts are derived. The upper end or *manubrium* (handle) starts at the base of the neck. The clavicles articulate with the manubrium. The cartilage of the first rib is attached to the manubrium. The *gladiolus* (body) of the sternum, like the manubrium, is a flat bone lying close under the skin.

The tip of the sternum is called the *xiphisternum* and is made of cartilage. The sternum has seven pairs of true ribs joined to it by cartilage strips.

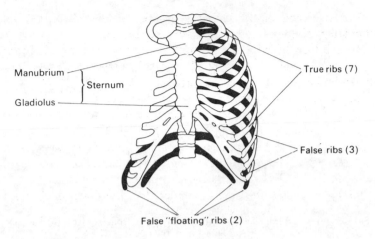

Manubrium

Sternum

Gladiolus

True ribs (7)

False ribs (3)

False "floating" ribs (2)

FIG. 21. The rib-cage.

The ribs. There are twelve pairs of ribs making up the thoracic rib-cage. These flat and curved bones articulate at the back with the twelve dorsal vertebrae. On the sternal side of the rib-cage some ribs join the sternum; others do not do so.

True ribs. There are seven pairs of true ribs. These are ribs which join the sternum by means of strips of cartilage.

False ribs. There are five pairs of false ribs. These are ribs which do not join the sternum. In the upper three pairs, the cartilages fuse with the cartilages of the ribs above. The last two pairs are free on the sternal side; they are known as "floating ribs".

The vertebral column. The backbone is made up of thirty-three irregular bones called *vertebrae*. Each of these bones is a modification of a "typical" vertebra.

A *typical vertebra* consists of the following parts:

A *body*, the weight-bearing part of the vertebra.

A *bony arch*, with a *neural canal* which encloses the spinal cord. The bony arch has three projections to which muscles attach, a *spinous* process and *two transverse* processes. Each vertebra has *articular surfaces* where one vertebra articulates with another, above and below.

Intervertebral discs are flexible pads between the bodies of the

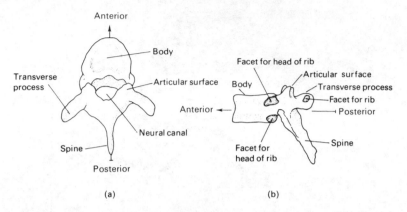

FIG. 22. (a) A typical vertebra (thoracic) seen from above. (b) Thoracic vertebra—from the side.

vertebrae, they make up about a quarter of the length of the backbone. These discs consist of a jelly-like core enclosed by tough fibrous tissue which is attached to the bodies of the vertebrae. Sometimes the weakest point of this disc may allow the jelly-core to bulge out and press on a nerve ("slipped disc").

The vertebrae may be divided into the following groups:

Cervical vertebrae	7 in number
Thoracic vertebrae	12 in number
Lumbar vertebrae	5 in number
Sacral vertebrae	5 in number
Coccygeal vertebrae	4 in number

The cervical vertebrae are the vertebrae of the neck. There are seven such vertebrae in the necks of all mammals. The first two vertebrae are modified to allow movements of the head.

The atlas is the first neck vertebra. It is a ring of bone with no body, but only a large neural canal. It has no spinous processes. This construction allows the head to nod.

The axis is the second cervical vertebra. This vertebra has the unmistakable *odontoid process* which projects upwards from the body of the vertebra into the neural canal of the atlas. This arrangement allows the head to rotate from side to side.

The thoracic vertebrae are the rib-carrying vertebrae and form a

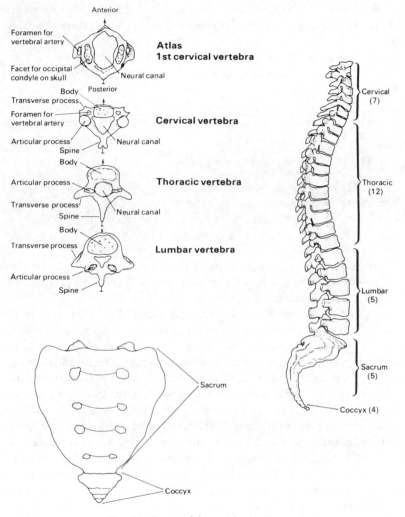

FIG. 23. Bones of the vertebral column.

backward curve down the thorax. They are characterized by their large spinous, processes and heart-shaped bodies. The ribs articulate with surfaces on the transverse processes and on the body.

The lumbar vertebrae are strong stout bones with no rib articulations. To these are attached the large muscles supporting the back. They are the largest bones in the vertebral column.

The sacral vertebrae are five fused vertebrae forming the *sacrum*. The sacrum articulates with the iliac bones of the hip bone to form the back of the pelvis.

The coccygeal vertebrae (or coccyx) are four vertebrae fused in adult life. They are fused with the sacrum above and form a vestigial tail.

(iv) The Pelvic Girdle and Lower Limbs

The *pelvic girdle* is the complete ring of bone formed by the lower part of the backbone and the two hip bones.

  (a) The vertebrae involved are those of the sacrum and coccyx. The Sacrum is attached on either side to the hip bones at the *sacro-iliac joints*.

  (b) The hip-bones are made up of three units.

   The *ilium* is the upper part of the pelvic bowl. It supports the sides of the abdomen.

   The *ischium* is the part upon which one sits.

   The *pubis* the forward projecting part of the pelvic girdle. It formed where the two halves of the pelvic girdle join at the *pubic symphysis*. The two pubic bones are bound together strongly by fibres in the male but less so in the female because they must allow movement during childbirth.

   All three hip bones go to make up the *hip-joint* or *acetabulum*, the socket into which the thigh bone fits. It is a ball and socket joint.

   The back wall of the pelvic bowl is formed by the sacrum which is joined to the iliac bones.

The lower limb consists of the following bones:

  The femur: thigh bone.

  The patella: knee cap.

  The tibia: shin bone ⎫ leg.
  The fibula ⎭

  The tarsals (tarsus): seven bones forming the ankle, heel and instep.

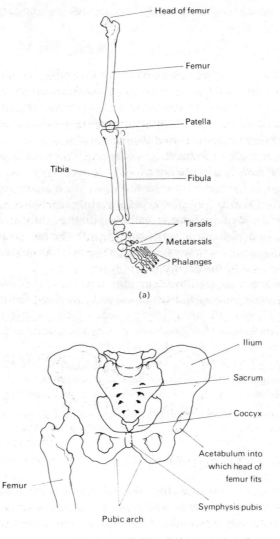

— Head of femur

— Femur

— Patella

Tibia —

— Fibula

Tarsals

Metatarsals

Phalanges

(a)

Ilium

Sacrum

Coccyx

Acetabulum into
which head of
femur fits

Femur —

Symphysis pubis

Pubic arch

(b)

FIG. 24. (a) Bones of the lower limb. (b) The pelvic girdle.

The metatarsals (metatarsus): five bones making up the sole of the foot.

The phalanges: three bones forming the small toes. Two bones form the big toe.

The femur is a strong, long bone with a hemi-spherical *head* which fits into the acetabulum of the pelvic girdle. Around the head are bony attachment areas called the *greater* and *lesser trochanters*. The lower end of the femur bone articulates with the tibia to form the knee joint.

The patella or knee cap is embedded in a long tendon which runs over the knee joint. The function of the patella is to ease movement of this tendon across the knee joint.

The tibia is the large bone in the leg, with a sharp forward crest forming the "shin-bone". The bone is slightly triangular in section.

The upper end of this bone articulates with the condyles of the femur. The lower end of the bone articulates with the bones of the ankle (the talus or astragalus). The tibia is attached to the fibula at the lower end, and runs down the inside of the leg.

The fibula is a thinner bone running down the outside of the leg. It joins the tibia just below the knee joint and just above the ankle.

The tarsus is made up of seven bones which are tightly united to form the ankle joint, heel and instep. The talus bone articulates with the tibia and fibula to form the ankle joint.

The metatarsus consists of five bones which articulate with the tarsus and phalanges to form the sole of the foot.

The phalanges are the small bones which make up the toes on the same *pentadactyl* plan as the hand.

SOME BONE DISEASES AND DISORDERS

Acromegaly. An upset of the pituitary gland can result in an excess of growth hormone being secreted into the blood. The bones of skull, face and hands, feet and internal organs increase in size in the adult. (In childhood excess of this hormone produces 'gigantism'.)

Fracture. A broken bone. If a bone breaks it usually snaps completely, but not so in young children, it is partial and known as a 'greenstick fracture'.

Osteomyelitis. This is an infection of the bone with an abscess forming in the bone marrow. The pus from the abscess has little chance of escape and so it is likely to move lengthwise along the bone shaft, thus spreading the infection. Some parts of the bone may die because the abscess interferes with the proper blood circulation. With no blood supply such areas of dead or dying bone are difficult to treat by blood borne drugs. Surgery may be necessary to open up the bone.

Protrusion of the intervertebral disc. Severe strain on the backbone by bending and lifting incorrectly may bring about a bulge on one side of an invertebral disc. This bulge may press onto some nerve root or other nerve fibres. This 'slipped-disc' disorder will show symptoms of backache and sciatica or lumbar pain.

Rickets. A bone deformity caused by a deficiency of Vitamin D in the diet, particularly in young children. Without Vitamin D not enough calcium salts are deposited in the bone to make it sufficiently rigid.

Talipes is the general name for various types of 'club-foot'. The foot does not lie at a normal angle to the leg; the heel may, for example, be drawn upwards.

CHAPTER 3

Joints and Muscles

THE movement of the skeleton is possible because of muscles attached to bone on either side of a joint. The function of the muscle is to contract and pull one bone towards another.

JOINTS

A *joint* is a place where two bones come close together. Joints may be classified according to their ability to move.

(a) *Fixed joints* are relatively immobile or completely immobile. These joints are bound closely together but do allow a little play in some areas such as the vertebral column where the tough joints are fibro-cartilage in nature. The bones of the skull are held together by fibrous tissue and are *not* capable of movement. Similarly, the pubic symphysis joint is slightly immobile, it is a fibro-cartilage junction which becomes looser in pregnancy to make birth possible.

(b) *Mobile joints* allow movement of one bone about another. A common example of a movable joint is the *synovial joint*. This joint is enclosed in a very tough fibrous capsule (ligament capsule). The inside of this capsule is lined by a synovial membrane which puts out a viscous lubricating and moistening synovial fluid. This synovial membrane does not lie over the articular cartilages which cap the two bones. It is possible to "pull a ligament" or "sprain" a joint which means the joint has been forced beyond the capacity of the ligaments to hold the two bones together. The ligaments may be damaged.

Synovial joints can be grouped as follows:

SKELETAL MUSCLES

Bones are moved at joints by the CONTRACTION and RELAXATION of MUSCLES attached to them.

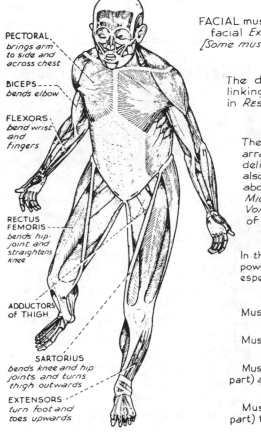

PECTORAL.
brings arm
to side and
across chest

BICEPS
bends elbow

FLEXORS
bend wrist
and
fingers

RECTUS
FEMORIS
bends hip-
joint and
straightens
knee

ADDUCTORS
of THIGH

SARTORIUS
bends knee and hip
joints and turns
thigh outwards

EXTENSORS
turn foot and
toes upwards

FACIAL muscles are involved in varying facial *Expression; Speech; Mastication* [*Some muscles link bone to skin.*]

The deep muscles of the THORAX, linking the ribs, contract and relax in *Respiration*.

The muscles of the ABDOMEN are arranged in sheets and *Protect* delicate abdominal organs. They also contract to compress abdominal contents and aid in *Micturition, Defaecation, Vomiting* (and in the processes of *Childbirth* in the female).

In the LEGS are found the most powerful muscles of the body – especially those acting on the hip-joint.

Muscles which bend a limb at a joint are called **FLEXORS**.

Muscles which straighten a limb at a joint are called **EXTENSORS**.

Muscles which move a limb (or other part) away from the midline are called **ABDUCTORS**.

Muscles which move a limb (or other part) towards the midline are called **ADDUCTORS**.

Joints and muscles. (Reproduced by permission from *Nurse's Illustrated Physiology* by McNaught and Callander, E. & S. Livingstone Ltd., Edinburgh.)

FIG. 25. Skeletal muscles.

SKELETAL MUSCLES

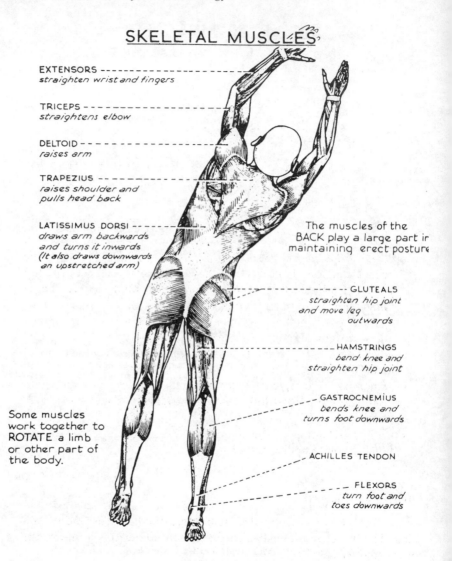

EXTENSORS - - - - - - - - - - - - - - - -
straighten wrist and fingers

TRICEPS - - - - - - - - - - - - - -
straightens elbow

DELTOID - - - - - - - - - - - - -
raises arm

TRAPEZIUS - - - - - - - - - - -
*raises shoulder and
pulls head back*

LATISSIMUS DORSI - - - - - - -
*draws arm backwards
and turns it inwards
(It also draws downwards
an upstretched arm)*

The muscles of the
BACK play a large part in
maintaining erect posture

- - - - - - - - - GLUTEALS
*straighten hip joint
and move leg
outwards*

- - - - - HAMSTRINGS
*bend knee and
straighten hip joint*

- GASTROCNEMIUS
*bends knee and
turns foot downwards*

Some muscles
work together to
ROTATE a limb
or other part of
the body.

ACHILLES TENDON

- - FLEXORS
*turn foot and
toes downwards*

FIG. 25.   (cont.)

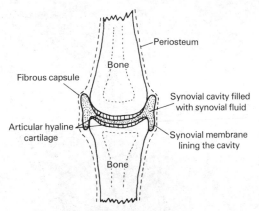

FIG. 26. A synovial joint in section (diagrammatic).

Ball and socket joints. Examples—hip and shoulder joints. The head of the long bone fits into the bowl-shaped cavity of the hip bone (acetabulum) or of the shoulder (glenoid cavity). These joints allow movements in many directions.

Hinge joints. Examples—elbow, knee and ankle joints. These joints allow movement in one direction only.

Double hinge joints. Example—wrist joint. This joint allows movement in two directions.

Gliding joints. Example--the articulating surfaces between vertebrae. This joint only allows a sliding movement of one surface over another.

Pivot joints. Example—the atlas on the axis. These joints allow one bone to rotate upon another.

JOINTS IN ACTION

(a) *The elbow* is a synovial joint between the humerus, radius and ulna. The muscles responsible for movement about this joint are the biceps and triceps:

The biceps lies in front of the humerus. It is attached by tendons at one end to the shoulder; at the other end, tendons fix the muscle to the radius bone, a little below the elbow joint.

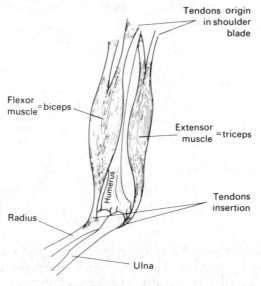

Tendons origin
in shoulder
blade

Flexor
muscle = biceps

Extensor
muscle = triceps

Humerus

Tendons
insertion

Radius

Ulna

FIG. 27. The elbow joint.

The triceps lies behind the humerus. Tendons attach the muscle at
one end to the shoulder, and at the other end to the ulna.

Bending the arm or lifting an object in the hand involves the following
action.

The biceps contracts and pulls the radius upwards. The triceps
relaxes as this occurs. A muscle which produces bending is known as a
flexor muscle.

Straightening the arm involves the following:

The triceps contracts and pulls the ulna downward. As the triceps
contracts so the biceps relaxes.

A muscle which causes a limb to straighten is known as an *extensor
muscle.* Flexors and extensors are *antagonistic,* in that one must relax
in order that the other may contract.

(b) *The knee* is another synovial joint which permits bending and
straightening of the leg.

The antagonistic muscles of the upper leg are:

    the quadriceps in front of the femur;
    the biceps femoris behind the femur.

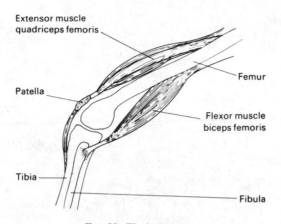

Extensor muscle
quadriceps femoris

Femur

Patella

Flexor muscle
biceps femoris

Tibia

Fibula

FIG. 28. The knee joint.

Mechanical movement in the human body is brought about by means of living leverage. The bones act as levers through which the power of the muscles is used. The joint of the bony levers is the *fulcrum*.

In the human body *levers* are of three types:

First order levers have the fulcrum situated between the effort that is moving the load:

(effort—fulcrum—load)

e.g. Nodding the head back and forth. Contraction of neck muscles (effort)—atlas (fulcrum)—head (load).

Second order levers have the load situated between the fulcrum and the effort:

(fulcrum—load—effort)

e.g. standing up and down on the toes. The ground (fulcrum)—body weight (load)—muscles raising the heel (effort).

Third order levers have the effort exerted between the fulcrum and load:

(fulcrum—effort—load)

e.g. bending the arm at the elbow. Elbow joint (fulcrum)—biceps contraction (effort)—object lifted (load).

The muscles that move these levers are attached at two points. These points of attachment are given special names.

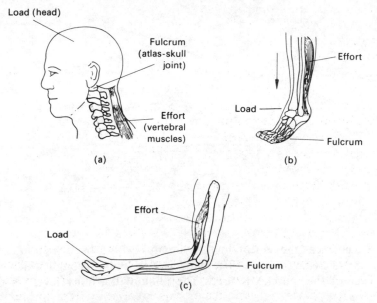

FIG. 29. Levers: (a) first-order lever; (b) second-order lever; (c) third-order lever.

The origin of a muscle is the region of tendon attachment on an immovable bone.

The insertion of a muscle is the region of tendon attachment on a movable bone.

The contraction of a muscle normally pulls the point of insertion toward the origin, e.g. the arm and biceps muscle:

Origin: in the shoulder girdle.

Insertion: a short distance below the joint at the elbow.

MUSCLES IN ACTION

·Muscles are very special tissues because they are capable of contraction when stimulated. This stimulation is normally a nerve impulse arriving at the muscle by way of a *myo-neural* (neuromuscular) junction.

Muscles are capable of behaving as they do because of their rather special chemical make-up.

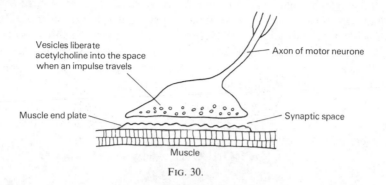

Vesicles liberate
acetylcholine into the space
when an impulse travels

Axon of motor neurone

Muscle end plate

Synaptic space

Muscle

FIG. 30.

(a) *The composition of a muscle* includes the following:

Actomyosin   a protein responsible for the contraction.

Myoglobin   a substance similar to haemoglobin in that it holds oxygen. Oxygen is required for muscle activity.

Phosphates   chemical materials involved in the transfer of energy necessary for contractions.

Glycogen   "animal starch" a store of glucose. Muscle activity needs energy. (Heart muscle gets its energy from fat rather than glucose).

(b) *Muscular contraction* involves the reaction together of two intertwined protein bundles—*actin* and *myosin*. When these two threads react together they form actomyosin. The result of this chemical reaction is that the muscle fibres contract. This happens when the muscle is stimulated. The energy needed for muscular contraction is obtained from a complex chemical called *adenosine triphosphate* (ATP). When a muscle is stimulated it triggers off the breakdown of this ATP to ADP (adenosine diphosphate) with the liberation of energy.

It is necessary to make more ATP as a store of energy otherwise it all gets used up and we could no longer use our muscles. The energy required to build new stores of ATP from ADP comes from the sugar (energy foods) in our diet. Glucose is stored in muscles as glycogen so here is a ready source of energy.

To make energy from glucose we need oxygen. Often we use our

muscles so quickly, for example in sprinting, that we cannot immediately get enough oxygen to the muscles. In this case we contract our
muscles in the absence of enough oxygen (*anaerobic respiration*), just
for a while, and build up an *oxygen debt*. We cannot do this too long
because an acid, lactic acid accumulates in the muscles causing *fatigue*.
We need a period of rest after this in order to get in enough oxygen to
oxidize, and get rid of the lactic acid (*aerobic respiration*).

$$C_6H_{12}O_6 \text{ (glucose)} \xrightarrow{\qquad} \nearrow \text{energy } 2CH_3CHOHCOOH \\ \rightarrow \text{(lactic acid)}$$

Anaerobic respiration

$$2\ CH_3CHOHCOOH + 6O_2 \rightarrow 6CO_2 + 6H_2O$$

Aerobic respiration

$$C_6H_{12}O_6 + 6O_2 \xrightarrow{\qquad} \nearrow \text{energy} \\ \rightarrow 6CO_2 + 6H_2O$$
glucose

Muscle tone. The muscles are normally in a state of slight tension,
which is called *"muscle tone"*. It keeps the structures to which the
muscles are attached in their correct positions. This muscle tone does
not produce fatigue because groups of muscle fibres alternate in their
contraction and relaxation. Muscle tone is maintained because the
muscles can adjust to incoming stimuli from receptors within the
muscles which detect movement and changes in position. *Good posture*
is maintained if no group of muscles is unnecessarily tense or contracted
in order to maintain balance or comfort. Fatigue will result in groups
of muscles under strain because of bad posture. Standing and sitting
should be as erect as possible, for this permits the body weight to be
distributed evenly upon the feet or buttocks. A curved back will put
strain on the low back muscles and will also restrict adequate breathing
movements. The abdominal contents are compressed if the spine and
shoulders curve forward.

SOME DISEASES AND DISORDERS OF JOINTS AND MUSCLES

Fibrositis. A blanket term for vague pain in and around muscles.
These pains seem to come and go without any apparent pattern.

Faulty posture has been suggested as one way to bring on these pains.

Gout. This is a painful acute disorder of a joint brought about by the precipitation of uric acid crystals at that joint. The classical example is the first joint of the big toe. It does not appear to relate exclusively to the love of alcohol as so often believed.

Muscular dystrophy. A defect of muscle fibres resulting in muscular wasting as though they have lost their nerve supply. The nerve supply to the muscles is, however, normal.

Osteoarthritis (Osteoarthrosis). This is a disorder of joints where the cartilage linings at the ends of bones degenerate making movement of those joints uncomfortable. The joints of the legs and spine seem to be most affected. The disorder is not an inflammatory one (hence the more appropriate word in brackets) but seems to be the results of "wear-and-tear" developing more commonly in middle-aged people.

Rheumatism. A general term used to describe any painful disorder of joints or muscles not related directly to an infection or injury.

Rheumatoid arthritis. A disease of the connective tissues characterized by bumps of inflamed fibrous tissue just beneath the skin. The knuckle and wrist joints are most commonly affected. Commoner in women than men.

Rheumatic fever. A disease with inflammation of various connective tissues. The tissues of larger joints and the valves and linings of the heart are commonly affected. It is most frequently found in younger people often following on from some bacterial infection. Damage to the heart valves may be a complication.

Tetanus. An infection with the tetanus bacillus, *Clostridium tetani*. It causes a toxin to be released into the tissues which follows up nerves and travels to the spinal cord. Convulsions, and spasma of voluntary muscle activity take place. A spasm of the jaw muscles gives rise to the name "lockjaw". Tetanus spores are quite often found in the soil.

CHAPTER 4

Nutrition

FOOD is any solid or liquid which, when swallowed, provides the human body with material enabling it to function in one or more of the following ways:

(a) The production of heat, or other forms of energy (fuel foods—carbohydrates and fats).

(b) The growth, repair or reproduction of body tissues (body-builders—proteins).

(c) Regulation of the production of energy or the processes of growth, repair and reproduction (vitamins—salts).

Substances which contribute to the above process are known as *nutrients*.

The study of nutrients, their structure and the way in which they are used by the body is within the study of *nutrition*. The study will here be considered under the following titles.

(a) *Foods*.

(b) *Diet* and energy requirements of the body.

(c) *Metabolism of nutrients*.

(d) *Digestion and the digestive organs*.

FOODS

All solids and liquids taken (ingested) into the body are not necessarily foods. A food contains substances which, when subjected to chemical action by the body become soluble, are absorbed into the blood and transported to areas where they have a specific function. The essential nutrients which are necessary for proper body functioning and which must be present in certain proportions in food are as follows:

Carbohydrates such as starch and sugar are fuel nutrients.

Proteins such as meat and eggs are tissue builders and repairers.

Fats such as mutton fat and vegetable oils are fuel and storage nutrients.

Mineral substances give structure to bone tissue. They assist the regulation of tissue activities. Salts of many types are grouped here.

Water. Two-thirds of the body is water. It is the basis of tissue fluids, digestive juices, urine and sweat.

Vitamins are essential regulators of tissue activity. Only minute quantities are necessary.

The incorrect balance of these nutrients produces a condition known as *malnutrition*.

Carbohydrates

These are complex organic chemicals, such as the sugars and starches, which are compounds of carbon, hydrogen and oxygen. The formula of a molecule of a simple sugar, such as glucose, will show that there are twice as many hydrogen atoms as there are oxygen atoms:

$$\text{Glucose} - C_6H_{12}O_6$$

All carbohydrate molecules are the same in so far as they have twice as many hydrogen atoms as oxygen atoms:

A general formula:

$$\text{Carbohydrates} - (C_xH_{2n}O_n)$$

x = number of carbon atoms, n = number of hydrogen and oxygen atoms.

The Sugars are grouped as:

Monosaccharides are simple sugars: glucose (grape sugar), fructose (fruit sugar, found in fruit juices and honey).

Disaccharides are complex sugars: sucrose (cane or beet sugar), lactose (milk sugar), maltose (found in beer).

The disaccharides consist of two molecules of a simple sugar united together, e.g.

$$sucrose = glucose + fructose.$$

The starches are water-insoluble carbohydrates, formed by the combination of many sugar molecules. Most plant carbohydrates stores are in the form of starch in granules. The capsule of these granules is ruptured when heat is applied to food containing such granules. Starch is the fuel food reserve of plants.

Glycogen is similar to starch, and is formed in the animal liver and muscle cells from molecules of glucose. Glycogen is the fuel food reserve of animals.

Cellulose is another (polysaccharide) complex carbohydrate formed from many simple sugar molecules. This substance forms the tough material from which plant walls are formed. In the animal world, carbohydrate is not used for structural purposes. Animals use carbohydrates because of the energy they release.

Cellulose is not digested by man.

The sources of some carbohydrates in food are listed as follows:

Carbohydrate	Source in food
Glucose	Fruits, plant leaves and roots
Fructose	Fruit juices, honey
Sucrose	Sugar cane, sugar beet
Lactose	Milk
Starches	Wheat, potato, rice
Celluloses	Green leaf vegetables, roots
Glycogen	Liver, muscle (meat)

THREE TESTS FOR CARBOHYDRATES

Benedict's test. One test for glucose present in urine employs Benedict's sugar reagent. A few drops of urine are added to 5 ml of the reagent and the solution is boiled for a minute or two. A green, yellow or red precipitate indicates glucose.

Fehlings test. The presence of a reducing sugar is shown by a red precipitate of cuprous oxide when solutions A and B are warmed with the suspect sugar.

Starch: Iodine test. Place a drop of iodine solution on a potato slice. A positive test for starch is the blue-black coloration produced with iodine.

Proteins

These are complex compounds of carbon, hydrogen, oxygen, nitrogen and sometimes sulphur. They make up about 12% of the weight of the human body.

Proteins are the basis of all animal and plant tissues since protoplasm is a protein material. They are the "body-builders". The molecules of proteins are large and must therefore be broken down into their smaller soluble components, the *amino-acids,* before being transported about the body.

Proteins are constructed of long chains of nitrogen compounds called *amino-acids.* About twenty-two different kinds of amino-acid are involved. The biological properties of a protein depend on the exact sequence of different amino-acids in the chain (primary structure), and on its orientation or shape as a whole.

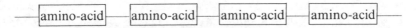

The proteins of food have to be broken down so that the amino-acids can be rearranged as human proteins.

The sources of some proteins in food are shown in Table 3.

Amino-acids are of different types: some are essential to the body, some are not.

The essential amino-acids must be present in man's food because the body is unable to synthesize them in sufficient quantities. A classification of proteins based upon the amino-acid content is:

First class proteins containing all or most of the essential amino-acids in good proportions, e.g. animal proteins in meat, eggs and milk.

Second class proteins containing only a few of the essential amino-acids, being thus poorer body-builders, e.g. plant proteins in beans, peas and cereals.

TABLE 3

	Source in food	Approx. percentage of protein (by weight)
Animal proteins		
Casein	Cheese	25
Myosin	Corned beef	25
Myosin	Fish	16
Myosin	Beef	13
Albumen	Egg	13
Casein	Milk (liquid)	3
Vegetable proteins		
Gluten	Soya flour	40
Legumin	Peanuts	28
Gluten	Flour	11
Legumin	Peas, beans	6
Legumin	Potato, cabbage	2

TWO TESTS FOR PROTEINS

Nitric acid test (Xanthoproteic test). A protein will form a yellow complex with concentrated nitric acid. Place a drop of concentrated nitric acid on coagulated egg white and a yellow coloration will be produced. An orange colour will form if ammonium hydroxide is added.

Millon's test. A protein boiled up with Millon's reagent will develop a pinkish-brown coloration.

Fats

Fats are compounds of carbon, hydrogen and oxygen. They are similar to carbohydrates, but contain less oxygen in proportion to the hydrogen. Fats, like the large molecules of starch and proteins, are insoluble in water. The fat molecule must be broken down into its component parts before it can be utilized by the tissues.

The fat molecule is a compound of glycerol (glycerin) and different fatty acids.

Beef fat	Glycerol—stearic acid.
Butter fat	Glycerol—butyric acid.
Olive oil	Glycerol—oleic acid.
Palm oil	Glycerol—palmitic acid.

Fats are an important energy store for the body. They yield twice as much energy as an equal mass of sugar. Fats alone are inadequate as a source of energy, as will be seen from the studies of fat metabolism.

The sources of fats in some foods are shown in Table 4.

TABLE 4

Food source	Approx. percentage of fat (by weight)
Frying oil or olive oil	100
Lard	99
Margarine	85
Butter	83
Almonds	54
Bacon	45
Cheese	35
Beef	28
Egg	12
Milk	4

THREE TESTS FOR FATS

Osmic acid test. When this dangerous acid is added to fats they become black.

Grease spot test. A fat or oil dissolved in a solvent, such as alcohol, or carbon-tetrachloride, will leave a grease smear when put on to a filter paper. This grease spot is translucent when the filter paper is held up to the light.

Sudan IV test. Fats, waxes and oils selectively take up this red stain.

Mineral Substances—Salts

Inorganic elements must be included in the diet, in order to give structure to bones and teeth and to maintain the composition and stability of body fluids and protoplasm. The salts of the following elements are very important for the correct functioning of the body.

Sodium, in the form of sodium chloride (common salt), is contained in all body fluids and is maintained at a definite concentration in times of health.

Any shortage of sodium chloride in the diet produces muscular cramps. People who work in hot temperatures tend to lose salt in their sweat and therefore require a diet with a good salt content. The amount of salt lost in the urine is controlled by the kidneys. The requirement of sodium per day is less than 1g if no sweating is occurring.

The carbonates and phosphates of sodium are of vital importance to the tissues and fluids of the body.

The sodium salt content in 28 g (1 oz) of some foods is given below:

Bacon	345 mg	Bread	117 mg
Cornflakes	298 mg	Egg	38 mg
Cheddar cheese	174 mg	Milk	14 mg

These figures and those following are taken from the *Manual of Nutrition* (HMSO).

Potassium salts are usually bound within the cell structure. Sodium salts are free in the body fluids. Potassium salts are not lost in the sweat. Plant materials are a common source of potassium.

Magnesium salts form part of the composition of bones and teeth. These salts are present in most foods.

Fluorine is found in different quantities as a fluoride salt in water. Small quantities of fluoride in water are thought to harden the teeth and thus to reduce the incidence of dental decay.

Phosphorus compounds in the body are generally phosphates. These phosphates are of vital importance in the chemical reactions which produce energy for muscular contraction.

Phosphates are also components of bones and teeth.

The amount of phosphorous compounds found in 28 g (1 oz) of some foods is listed below:

Cheese	155 mg	Eggs	62 mg
Liver	89 mg	Milk	27 mg
Kidney	74 mg	Fruit	5 mg

Iron is an element around which the molecule of haemoglobin (red blood pigment) is built. Deficiency of iron may result in anaemia. Iron salts are administered in cases of anaemia, but iron deficiency is only one of its causes.

Iron compounds are found in the quantities listed below in 28 g (1 oz) of the foods indicated.

Liver, kidney	3·9 mg	Egg	0·9 mg
Beef (corned)	3·1 mg	Potato	0·2 mg

Calcium salts are necessary in our diet in order to ensure:
(a) The ossification of bones and teeth.
(b) The regulation of the excitability of nerve fibres and nerve centres.
(c) The continued contraction of the heart.
(d) The clotting of blood and the curdling of milk in the stomach.

Calcium salts are particularly important in the diets of young children, expectant and nursing mothers and old people.

Calcium deficiency results in poorly developed bones and, eventually, rickets in children. A shortage of calcium in the diet of old people may cause them to have brittle bones. The quantity of calcium compounds in 28g (1 oz) of some foods is indicated below:

Cheese	230 mg	Egg	17 mg
Sardines	113 mg	Meat	3 mg
Milk	34 mg	Potato	2 mg
Cabbage	18 mg		

Iodine is important for the formation of thyroxine, the secretion manufactured by the thyroid gland. Only small amounts of iodine are required in the diet. Deficiency of iodine will result in thyroid gland disorders. Iodine is found in watercress, onions, sea food, liquid milk and most drinking water.

Water

The daily intake of water is the most important of our dietary requirements. It makes up about 60% of man's body weight. This weight remains fairly constant from day to day, suggesting our water content remains constant. Water lost is normally balanced by water gained. Someone has calculated that the body loses about 2·0 litres of water daily, about 1·0 litre in the urine, 600 cm^3 by evaporation from the skin, 300 cm^3 from the lungs and about 100 cm^3 in the faeces. This must be replaced by eating and drinking.

Other authorities suggest our daily loss of water is twice the above figure, but whatever the case, obvious body changes become noticeable when about 10% water loss is experienced, if water is not taken in to counteract the loss.

A man can only survive a few days without drinking because eventually the blood becomes more concentrated and viscous and the blood pressure may drop to a point that circulation ceases.

Vitamins

Vitamins are vital complex chemical substances, which in minute quantities are essential to the diet. Their function is regulatory.

The need for some chemical ingredient in the diet, in addition to the foregoing nutrients, was discovered in the early part of the twentieth century. The deficiency of such chemical nutrients was seen to result in disease.

Vitamins may be grouped in terms of their solubility in fats and water. Each vitamin as it was discovered was given a letter. This is not a very good system of classification because many of the original vitamins have been discovered to be groups of compounds. The use of a chemical name is better.

Many vitamins are *biological catalysts* (see Glossary).

(a) FAT-SOLUBLE VITAMINS A, D, E AND K

Vitamin A (retinene or retinol) is a long-chained alcohol found in fish oils, butter and meat. It is also found in eggs and liver fat, particularly in the liver oils of cod and halibut. Two molecules of vitamin A joined together produce the substance called *β carotene* which is found in green vegetables and carrots. They are converted into vitamin A in the body.

Vitamin A deficiency leads to changes in the epithelia.
  (a) Mucous membranes become hard and dry (stratified).
  (b) The conjunctiva of the eye becomes opaque (cornified) causing a condition called *xerophthalmia*.
  (c) The sweat glands, the tear glands and salivary glands degenerate.
  (d) "Night blindness" is one of the earliest symptoms of deficiency, because vitamin A is essential for the regeneration of *visual purple* in the eye, which is bleached when light falls upon the retina.

(e) Degeneration of tooth enamel is another symptom.

Some foods with decreasing vitamin A content are listed below:

Animal sources	*Vegetable sources*
Halibut liver oil (larger quantities)	Carrots (larger quantities)
Cod liver oil	Spinach
Butter	Tomato
Cheddar cheese	Cabbage (small quantities)
Egg	
Meat (small quantities)	

Vitamin D (calciferol) is a vitamin produced in the skin by the action of sunlight; it prevents a bone deformity called *rickets*. This vitamin is taken by mouth in our diets as cod-liver oil, milk, egg yolk.

Vitamin D has been shown to be two substances, D_2 and D_3, *calciferol* produced in the skin by ultra-violet radiation and *cholecalciferol* found in fish liver oils.

Vitamin D deficiency results in the following disorders:

(a) Poor development of bones and teeth because of calcium loss across the intestine and in the urine.

(b) In severe cases *rickets* can result. This is a very bad form of deformity resulting from poorly developed and "soft" bones.

Some foods with decreasing vitamin D content are listed below:

Cod-liver oil (larger quantities)	Egg
Sardines	Butter
Herring	Milk (small quantities)

Vitamin E (Tocopherol) is found in large amounts in wheat-germ and cotton-seed oil.

The absence of this vitamin causes sterility in laboratory rats. The female conceives but does not produce live offspring.

There is no evidence of this vitamin being essential to man.

Vitamin K is found in green vegetables, cabbage, sprouts and spinach.

It is essential for the formation of prothrombin by the liver. Prothrombin is important in the sequence of events that leads to blood-clotting.

A deficiency of this vitamin can increase the possibilities of haemorrhages.

Vitamin K requires the presence of bile in the intestine in order to be absorbed by the gut wall.

WATER-SOLUBLE VITAMINS—B COMPLEX AND C

Vitamin B complex is a complex of at least a dozen different substances, often found in the same foods. They are involved in the energy-producing mechanics of the body for which oxygen is necessary. Some are involved in the production of the important nucleic acids found in the nucleus and other places.

Vitamin B (Aneurine or Thiamine). This is found in yeast and in the germ of cereals. It is sometimes known as the *anti-beri-beri factor* because it prevents the onset of a rather unpleasant disorder of the nerves and, eventually, the heart. It is needed in the diet so that we can use carbohydrates properly.

This vitamin is found in the foods listed below; they are arranged with best sources at the top of the list.

Brewer's yeast (largish quantities)	Potato
Peanuts	Cabbage
Bacon	Milk (smallish quantities)
Peas	Beer is a good source of this
Mutton	vitamin

Vitamin B$_2$ (Riboflavin). This vitamin is found in wholemeal flour, milk, meat, yeast, tomatoes.

Deficiency of this vitamin shows itself in skin disorders with inflammation of the mouth and tongue as examples. Misting over of the eyes may develop.

This vitamin is found in the foods listed below:

Brewer's yeast (largish quantities)	Egg
Liver	Beef
Cheese	Milk (smallish quantities)

Vitamins which are also involved as enzymes in the important energy producing cycles (see internal respiration later—p. 89) are the following two B complex vitamins.

Nicotinic acid (Niacin or Nicotinamide). This vitamin is found in liver, kidney and yeast. It is sometimes known as the *anti-pellegra* factor because its inclusion in the diet prevents the onset of pellegra. This disease produces an inflamed and blistered skin, chronic diarrhoea and eventual disturbance of nerve and brain function. People living on poor diets of which maize is the main component are liable to this disease.

Pantothenic acid. This vitamin, together with riboflavin and nicotinic acid, is involved as a catalyst in the important chemical reactions which produce energy.

Vitamin B$_6$ (Pyroxidine). A vitamin found in all plant and animal tissue. It is important in the chemistry that enables us to use fat and amino-acids.

Vitamin B$_{12}$ (Cyanocobalamin). A vitamin found in liver and important in the prevention of *pernicious anaemia*.

Folic acid is a vitamin factor which is important for the production of nucleic acids and the growth of the foetus. It is found in leaf vegetables, yeast, liver and kidney.

Vitamin C (Ascorbic acid). This water-soluble vitamin is found in fresh fruit such as blackcurrant, orange and grapefruit. It is also found in fresh vegetables such as cabbage, sprouts and tomatoes. It is not present in milk after pasteurization. This vitamin is easily destroyed by cooking at high temperatures.

Deficiency of this vitamin results in *scurvy*. This disease shows weakening of blood capillaries and the resultant haemorrhages beneath the skin. The skin heals very slowly if damaged. Gums bleed around the teeth. The lack of this vitamin tends to weaken the "cement" holding body cells together. Foods containing this vitamin are listed below:

Blackcurrants (largish quantities)	Lemons
Brussels sprouts	Lettuce
Cabbage, cauliflower	Onions, carrots (smaller
Oranges	quantities)

DIETS AND ENERGY REQUIREMENTS

Diet

A healthy body is maintained if the food intake is correctly proportion and sufficient in quantity. A *balanced diet* will contain:

Carbohydrates	roughly 4 parts
Proteins	roughly 1 part
Fats	roughly 1 part
Mineral salts	small proportions
Vitamins	traces

Water   variable
Roughage (cellulose of plants)   variable

Roughage is indigestible plant fibre material, which stimulates the lower bowel to empty, from time to time.

An adequate meal containing the above nutrients may be made up from the foods below:

Meat or fish	Green vegetables
Eggs	Potatoes
Fresh fruit	Bread and butter
Milk and cheese	

Milk from the cow is often regarded as a complete food. It compares favourably with human milk and is therefore used to feed human babies, when the mother for some reason does not do so. The milk must be treated by heating to a high temperature and then cooling in order to kill off harmful bacteria.

Pasteurized milk is milk heated to a temperature between 63°C and 65°C for half an hour or to 72°C for at least 15 seconds. The milk is then cooled down to a temperature not higher than 10°C.

Tuberculin-tested milk is milk obtained from cows specially licensed as being healthy and free from tuberculosis infection. T.T. milk may be pasteurized or sterilized to make it particularly safe.

Sterilized milk is filtered, homogenized and then heated in bottles to a temperature of 100°C for a given period of time. This milk should be completely germ free and unaltered in taste. It will, however, contain no vitamin C as it is destroyed at boiling point.

Human babies can survive on milk for several months but must (say by 4–6 months) be put onto solid foods in order to provide sufficient vitamins and mineral salts.

Mother provides a special milk for a few days after the birth, *colostrum*. This milk contains 8 % protein, 2·5 % fat and 3·5 % sugar. Normal milk has 2 %, 4 % and 8 % of these foods. The mother's milk depends upon her diet, in terms of vitamins, but the other components are supplied from her body tissues.

Energy Requirements

Energy is required by the body to maintain its everyday activities such as walking, standing, sitting, etc. Energy is also necessary for special activities of work or play. At rest the body requires energy for all the living processes such as excretion, digestion, circulation and so forth. These basal requirements at rest are determined by the size of the person and are closely related to the area of the body surface.

The *basal metabolic rate* (BMR) is the energy requirements under basal conditions measured as Calories per square metre surface area per hour.

The *Calorie** is the unit employed (a capital C).

The Calorie is the amount of heat required to raise the temperature of 1000 g (1 kg) of water by 1°C.

For man   the basal requirements are 40 Cal per square metre per hour.

For woman   the basal requirements are 37 Cal per square metre per hour.

An example. It can be calculated by a formula that a man with a surface area of 1·8 square metres (10 stone 5 lb=65 kg, 5 ft 8 in.= 174 cm) needs 1728 Calories per day.

The BMR is not only proportional to surface area of the body but also varies with sex and age. For example, energy requirements decrease with age.

8-year-old boy —51·8 Cal/m²/hour

16-year-old boy —45·7 Cal/m²/hour

60-year-old man—30·0 Cal/m²/hour

Over and above the factors of size, age and sex, additional energy is required for various physical activities. In this way it is possible to arrive at an idea of our daily total energy requirements.

An example

B M R	1728 Cal/day
Extras: Slow walking	140 Cal/hour
Dish washing	70 Cal/hour
Brain work	0 Cal/hour
Digging	120 Cal/hour
Walking upstairs	1000 Cal/hour

*Calories are not included within the S.I. units system [1 Cal. (15°C)=4185·5 joules].

If we say the extra energy required per day is 1000 Calories then the daily requirements are 2728 Cal/per day (i.e. BMR 1728 + 1000).
The energy required is obtained from food.

> 1 g of carbohydrate gives 4 Calories
> 1 g of protein   gives 4 Calories
> 1 g of fat   gives 9 Calories

The energy supplied in a diet, as below, would be enough to supply the energy for our previous example.

> 375 g carbohydrate — 1500 Cal
> 100 g protein   — 400 Cal
> 100 g fat   — 900 Cal
>
> Total energy   2800 Cal

Food and Fatness

As seen previously a person requires enough energy for chemical "tick-over" (BMR) and additional energy for physical activities. If more energy foods are eaten than required they are converted into storage fat. This is a tendency more obvious in some people than others, it may run in families.

Higher calorie foods are the more fattening.

In 2·8 g of the following foods, they produce the energy shown:

Fat, lard	253 Cal	Bread	72 Cal
Butter	211 ,,	Potato	21 ,,
Bacon	128 ,,	Milk	17 ,,
Cheese	117 ,,	Apple	12 ,,
Sugar	108 ,,	Beer	10 ,,
Beef	88 ,,	Cabbage	7 ,,

> A pint of milk will supply   340 Cal
> A pint of beer will supply   200 Cal

Slimming must involve a reduction in the calorific value of the food intake without upsetting the balance of the food nutrients eaten. Exercise must accompany any slimming process.

It is worth noting that "vibrating" fat by massage or heating fat in steam-baths, and so forth, have no effect upon the slimming process by removing fat.

DIGESTION AND THE DIGESTIVE ORGANS

Digestion

This is a chemical process which changes complex chemical units into more simple soluble units. These smaller molecules are more easily absorbed through the lining of the intestine into the bloodstream. The breakdown of these food chemicals is brought about by biochemical *catalysts* called *enzymes*.

A catalyst is a chemical substance which accelerates or retards a chemical reaction without itself being changed.

Enzymes are special proteins and are present in the digestive juices which are secreted onto the food in the earlier parts of the alimentary canal.

Enzyme activity in digestion is influenced by the following factors:

(a) Each enzyme has an optimum temperature for its activity. Excessive heat destroys their activity.

(b) Each enzyme has an optimum range of acidity or alkalinity for its activity.

(c) Each enzyme acts upon a specific food or substrate.
For example:

> Amylases act upon carbohydrates
> Proteases act upon proteins
> Lipases act upon fats

Most enzyme names are written with the suffix-*ase*.

The end results of digestive action are as follows:

Carbohydrates ⟶ amylases ⟶ glucose (pass into the blood)

Proteins ⟶ proteases ⟶ amino-acids (pass into the blood)

Fats ⟶ lipases ⟶ glycerol and fatty acids (pass into the lymph)

The Digestion of Carbohydrates

Carbohydrates, such as monosaccharide, glucose and fructose do not require digestion, they can be absorbed into the blood.

Complex carbohydrates, such as sucrose (dissaccharide) and starches (polysaccharides), need to be broken down before they can be absorbed. They are broken down by *amylase* enzymes into simple sugars such as glucose and fructose. These are then absorbed.

Some complex carbohydrates are useless to man because he has no enzymes to break them down. Cellulose cannot be digested by man because he has no cellulase enzyme.

TABLE 5

Location	Digestive juice	Digestive glands	Enzyme	Action/End-products
Mouth	Saliva	Salivary glands	Salivary amylase (ptyalin)	Cooked starch→maltose
Stomach	NO CARBOHYDRATE DIGESTION			
Small intestine	Pancreatic juice	Pancreas	Pancreatic amylase	Starch ⟶ maltose
	Intestinal juice	Intestinal glands	Intestinal Maltase Sucrase Lactase	Maltose ⟨ glucose + glucose; Sucrose ⟨ glucose + fructose; Lactose ⟨ glucose + galactose

The simple sugars produced as a result of carbohydrate digestion, glucose, fructose and galactose are taken up by the cells of the small intestine by an *active* process requiring energy. Fructose and galactose are later converted to glucose. Glucose is blood sugar, the excesses of which are stored in the liver and muscle as a polysaccharide, *glycogen*. It is the hormone *insulin* from the pancreas which brings about this conversion of glucose into *the insoluble* storage glycogen.

The normal blood glucose level is 100 mg glucose per 100 cm³ of blood. Levels of glucose in excess of this may pass over into the urine, in *diabetes melitus* (sugar diabetes).

The digestion of proteins

Proteins are made up from amino-acids. Long chains of the twenty or so amino-acids in different combinations or spatial arrangement make up all our body proteins. These chains do not necessarily contain all the amino-acids. The amino-acids needed to build up proteins are taken in the food. Some of these amino-acids are known as *essential* because the body is unable to make these acids in sufficient quantities and they must therefore be included in the diet. The other twelve or so acids can be made by the body if they are not included in the diet. Animal protein contains all the essential amino-acids.

Milk is a good source of essential amino-acids.

Food proteins are broken down into their constituent amino-acids by *protease* enzymes. The linkages joining amino-acids in the long protein chains are called *peptide linkages,* it is these linkages that are ruptured as protein is digested. Progressively shorter lengths of broken protein chain are produced at each stage of digestion by the enzyme action.

Protein-digesting enzymes are present in the digestive juices as *precursors,* that is as inactive forms because the stomach and intestinal linings are protein and in danger of themselves being digested. *Pepsinogen* becomes pepsin when the correct acid conditions are present. Stomach acid is *hydrochloric acid,* produced by *parietal* (or *oxyntic*) cells in the stomach lining. *Trypsinogen* becomes trypsin when it comes into contact with the intestinal enzyme activator, *enterokinase.*

The amino-acids produced as a result of digestion are absorbed from the small intestine into the blood.

Amino-acids circulating in the blood are used by cells for protein construction. Some amino-acids not used by cells for protein synthesis pass to the liver where they are broken down (amino-acid deamination). The ammonia produced as a result of this "nitrogen" removal process is converted to urea in the liver. Ammonia is very poisonous, urea is not.

TABLE 6

Location	Digestive juice	Digestive glands	Enzyme	Action/End product
Mouth	NO PROTEIN DIGESTION			
Stomach	Gastric juice	Gastric glands— chief cells	Rennin (young mammals)	Coagulates milk protein Caseinogen → Casein (soluble) (insoluble)
			Gastric protease (pepsin)	Proteins→proteoses+ peptones (longer peptide chains)
Small intestine	Pancreatic juice	Pancreas	Pancreatic protease (trypsin)	Proteoses→polypeptides+ peptones+amino-acids (shorter peptide chains)
	Intestinal juice (succus entericus)	Intestinal glands (crypts of Lieberkühn)	Intestinal peptidases (erepsin) enterokinase	Polypeptides→amino-acids An activator of trypsinogen (trypsinogen→trypsin)

People living on maize as their main diet will show a protein deficiency disorder called *Kwashiorkor*. Children with this disorder have distended abdomens because of swollen liver and retained water (oedema). They have little natural defence against disease. The distribution of dried milk has helped in some areas.

The Digestion of Fats

Fats are made up of glycerol and fatty acids. The digestion of fats and oils offers some problems because they are not soluble in the watery fluids of the body. In order to render fats available for digestion they are *emulsified* by bile salts. This detergent action keeps the fat droplets in suspension within the water.

Fat-splitting enzymes, *lipases,* produce fatty-acids and glycerol which are absorbed by the intenstinal mucosa lining. Within the intestinal cells neutral fat is produced by the resynthesis of glycerol and fatty acids. This neutral fat is seen as minute droplets within the central lacteal of the villus after a fatty meal. Later this fat enters the venous system by way of the thoracic duct (see p. 117).

Some shorter chains of fatty acid are taken up into the blood across the intestine and pass to the liver by way of the portal vein.

Fat is stored beneath the skin or within the abdomen in *fat depots.*

TABLE 7

Location	Digestive juice	Digestive glands	Enzyme	Action/End-product
Mouth	NO FAT DIGESTION			A little melting
Stomach	NO FAT DIGESTION			Mixing and melting
Small intestine	Pancreatic juice	Pancreas	Pancreatic lipase	Fats ⟶ glycerol + fatty acids
	Intestinal juice	Intestinal glands	Intestinal lipase	Fats ⟶ glycerol + fatty acids

Fat digestion requires an alkaline medium which is supplied by the salts within pancreatic juice. Bile juice, stored in the *gall-bladder,* and made in the liver, is put into the duodenum when fat enters it.

A jaundiced condition is when bile is prevented from entering the duodenum and results in vomiting when fat enters the small intestine.

The Digestive Organs

Food is taken into the human body at the mouth (ingestion). It is chewed (masticated), moved around by the tongue and mixed with saliva. It becomes a pellet called a *bolus.* The bolus is pushed backwards into the pharynx (throat) by a voluntary action, *swallowing.* From now onwards the muscular movements of the *alimentary canal* are involuntary *(peristaltic action).*

Food being swallowed is prevented from entering the nasal cavity by a sphincter near the *uvula* and from entering the larynx (voice-box) by the *epiglottis*. These "flaps" cover the entrances to the respiratory regions and so prevent choking or suffocation by food.

Once swallowed food is set on its way through the alimentary canal. The alimentary canal has digestive glands associated with it, they supply the fluids necessary to break down foods. The lining of the alimentary canal is a tissue able to absorb soluble foods and it generally secretes a lubricating mucus.

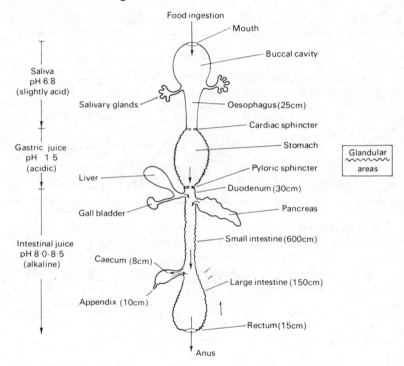

FIG. 31. The alimentary canal (diagrammatic).

THE MOUTH

The mouth is a cavity, bounded laterally by the cheek muscles. The floor of the cavity contains the tongue and teeth. The roof of the cavity is a bony palate, which becomes a soft palate towards the throat. The *uvula* is the name given to part of the soft palate structure which hangs at

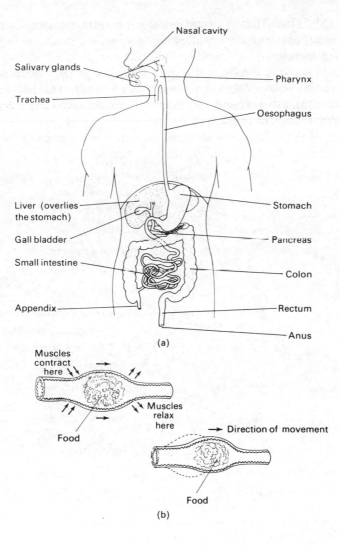

Nasal cavity

Salivary glands

Pharynx

Trachea

Oesophagus

Liver (overlies
the stomach)

Stomach

Gall bladder

Pancreas

Small intestine

Colon

Appendix

Rectum

Anus

(a)

Muscles
contract
here

Muscles
relax
here

Food

Direction of movement

Food

(b)

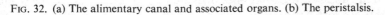

FIG. 32. (a) The alimentary canal and associated organs. (b) The peristalsis.

the back of the mouth. This soft palate prevents the entrance of food into the nasal cavity during swallowing, and assists in sound formation during speech.

The tongue is a muscular organ which assists in the mixing and swallowing of food. The mucous membrane of the tongue bears *papillae* which contain *taste buds*. The different sensations of taste are specific to certain areas of the tongue.

The tongue also assists in sound formation during speech.

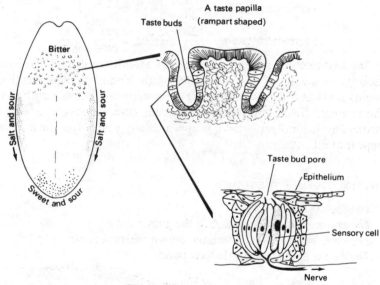

FIG. 33. The tongue and taste buds.

The teeth are hard bony structures used in *masticating* the food into small acceptable particles for reception by the alimentary canal. There are two sets of teeth developed in the human skull.

Temporary (milk) teeth: 20 teeth in childhood.

Permanent teeth: 32 teeth in adulthood.

The teeth present in the skull are grouped according to shape, position and function.

The incisors are chisel shaped cutting teeth situated at the front of the upper and lower jaws.

The canines are pointed teeth situated laterally next to the incisors.
The pre-molars are grinding teeth next to the canines.
The molars are the back grinding teeth.
The layout of the teeth in the skull is represented by the *dental formula*. This formula gives the number and type of teeth present in one half of the upper and lower jaw.

The dental formulae of temporary and permanent dentition are shown as follows.

$$\text{Temporary teeth} - \frac{2\,i}{2\,i} : \frac{1\,c}{1\,c} : \frac{2\,m}{2\,m}$$

$$\text{Permanent teeth} - \frac{2\,i}{2\,i} : \frac{1\,c}{1\,c} : \frac{2\,pm}{2\,pm} : \frac{3\,m}{3\,m}$$

The first teeth to appear in the child's mouth are usually the lower incisors, at the age of about 6 or 7 months. The permanent dentition begins at about 6 years when the 1st molars appear. The last milk teeth, the canines, are finally replaced by permanent canines at about 12 years. The 3rd molars (wisdom) appear many years later, or may not appear at all.

THE STRUCTURE OF THE TOOTH

The tooth consists of:
The crown is the portion above the gum.
The neck is the region where the crown meets the root.
The root is embedded in the jaw bone.

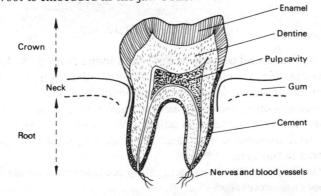

FIG. 34. Section of molar tooth.

The greater part of the tooth is constructed from a bone-like *dentine* which is covered by an even harder material called *enamel*. The root is covered by a *cement*. The inside of the tooth is hollow and contains a soft connective tissue, together with blood vessels and nerves. This is called the *pulp cavity*. Dental *caries* are produced when the enamel of the tooth becomes broken down by acids, and infection produces a degeneration within the pulp cavity.

The salivary glands produce a watery saliva which is poured into the mouth by way of ducts. There are three pairs of salivary glands.

The parotid glands are just in front of and below the ears.

The submandibular glands are situated inside the angles of the jaws.

The sublingual glands lie beneath the tongue.

The saliva is a digestive secretion poured into the mouth as an automatic (reflex) reaction to stimulation by certain odours, sights or thoughts. The presence of an object in the mouth will cause saliva to be secreted.

The composition of saliva is as follows:

Water which contains salts giving the saliva a slightly acid reaction— pH 6·35–6·85 (mixed saliva almost neutral to litmus). Water makes up about 99% of saliva.

Salivary amylase (ptyalin) is the enzyme which acts on cooked starch to produce maltose (malt sugar).

Mucus (mucin) is a lubricating substance allowing an easier passage of food.

About 1·5 litres of saliva is produced per day.

THE PHARYNX

This is the muscular back wall of the mouth, nose and throat. It extends as far as the opening of the gullet (oesophagus).

The pharynx side walls contain *lymphoid tissue* called *tonsils*. At the back of the pharynx, the *nasopharynx,* there is lymphoid tissue called the *adenoids*. This tissue sets up resistance to incoming germs (bacteria). They can themselves become infected.

The middle ear communicates with the nasopharynx by way of the *eustachian tubes* which open here. Equalization of air pressures on either side of the ear drum is permitted as this tube opens during swallowing.

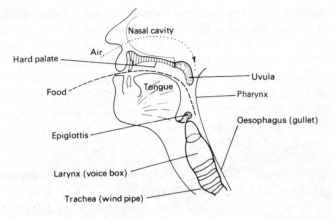

FIG. 35. The pharynx.

When swallowing takes place the *uvula* and *epiglottis* close over the respiratory passages to prevent choking.

THE OESOPHAGUS

This is a continuation of the pharynx and leads the food into the stomach. It is a muscular tube about 22–25 cm long, lined by mucous membrane. It passes down behind the trachea and enters the stomach through the cardiac orifice. Food which enters the oesophagus as a result of swallowing is carried to the stomach by involuntary waves of contraction in the oesophagus.

THE STOMACH

The stomach is a curved muscular bag, lying on the left of the abdomen just beneath the diaphragm (the spleen curves around the lower surface of the stomach).

The entrance to the stomach has a *cardiac sphincter*, a muscular ring through which food passes into the *fundic* region of the stomach. Air trapped in this fundic region can act as a resonator producing the stomach "gurgling" of the hungry when juices pass into the stomach from the oesophagus.

Below the stomach is the short tubular *pylorus* leading to the first part of the small intestine called the *duodenum*.

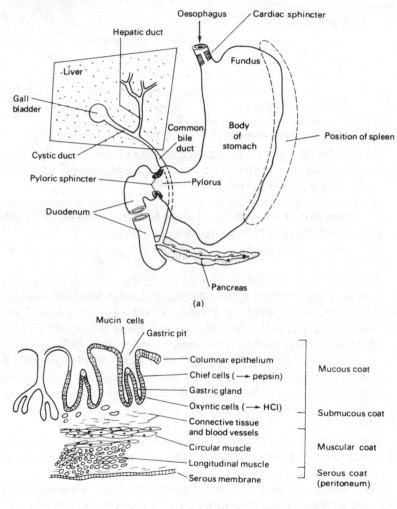

(a)

(b)

FIG. 36. (a) The stomach and associated organs. (b) Section across the stomach wall.

The stomach is lined by a folded mucous membrane ("tripe").

In section the stomach wall shows the following layers which have been illustrated previously.

An outer serous membrane known as the *peritoneum.*

A *muscular coat* containing fibres running in longitudinal, circular and oblique paths.

A *sub-mucous coat* of connective tissue and blood vessels.

A *mucous membrane* containing *gastric glands.* Mucus is secreted to assist lubrication.

The functions of the stomach are:

(a) To receive, mix and soften food. To melt fats.

(b) To absorb some water and alcohol, and some drugs.

(c) To commence protein digestion by secreting gastric juice containing a protease enzyme.

(d) To produce and release the *intrinsic factor* which encourages the absorption of vitamin B_{12} (known as the *extrinsic factor* being in food) thereby preventing pernicious anaemia.

The gastric juice contains:

Water

Hydrochloric acid which is secreted by the *oxyntic* cells in the walls of the gastric tubular glands. This acid is about N/10 (decinormal) 0·5 % and has a pH of 1·5–2·5 when food is in the stomach.

This acid kills many bacteria that are contained in the food.

Protease enzyme (pepsin) is presented in the inactive form, *pepsinogen.* It is produced by the *chief cells* (peptic cells) in the gastric glands. Pepsinogen converts to the active pepsin when hydrochloric acid is present in the stomach. Pepsin breaks down longer chains of proteins to smaller molecules.

Mucin

The stomach lining is prevented from attack by the pepsin by the secretion of mucus which covers the mucosa. Erosion of the lining may result in a *peptic ulcer.*

Rennin is an enzyme present in the gastric juice of young milk-drinking mammals. It curdles milk. It converts the soluble milk protein *caseinogen* to the insoluble milk protein *casein.*

Chemical or mechanical stimulation of the pyloric area of the stomach lining causes it to secrete a hormone *gastrin.* This hormone controls the production of gastric juice.

THE SMALL INTESTINE

The stomach pours a semi-liquid *(chyme)* into the curved duodenum, into which the pancreas secretes an alkaline pancreatic juice. Bile enters the duodenum also.

The duodenum is the first part of the small intestine. It is about 30 cm long.

The jejunum is the name given to the first 240 cm or so of the small intestine. It passes into the last coils of the intestine called the ileum.

The ileum is about 360 cm in length.

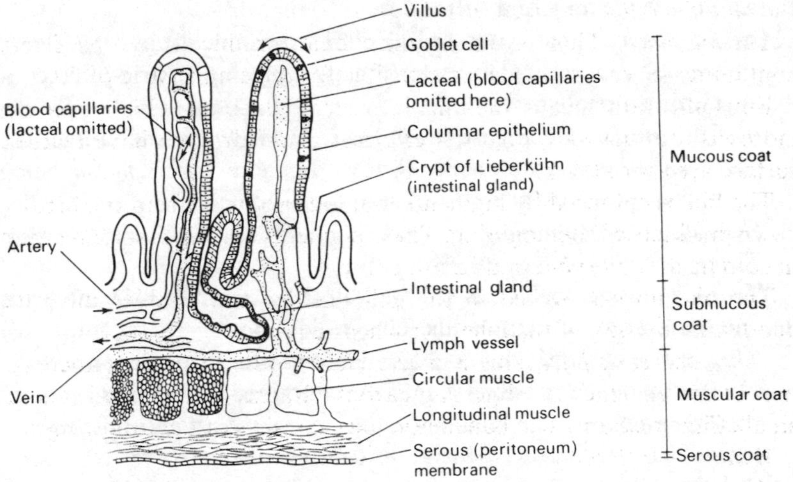

FIG. 37. Section across the small intestine.

The small intestine is held in coils in the abdomen by folds of the peritoneum called the *mesentery*. A transverse section across the small intestine shows the structure of the intestinal wall.

The small intestine wall has the same coats as the stomach wall:

An outer serous coat (or peritoneum).

A muscular coat containing longitudinal and circular muscles.

A sub-mucous coat consisting of connective tissue and blood vessels.

A mucous membrane which is thrown into folds. Minute finger-like
    villi cover the surface of this membrane. Each villus contains a

network of blood and lymph capillaries. Glucose and amino-acids are absorbed into the blood capillaries. Fat droplets pass into the lymph vessels or *lacteals*. The intestinal juice (succus entericus) is secreted by glands in the mucous membrane. Also occurring in the mucous membrane are lymphatic tissues which perform a "germ" filtering function. The grouping of this tissue in the ileum is known as *Peyer's patches*.

The functions of the small intestine are summarized into the following three categories.

(a) It receives the acid chyme from the stomach and neutralizes it by the *alkaline bile* and *pancreatic juices*.

The bile juice. This is an alkaline fluid produced by the liver, containing no enzymes. It is responsible for the emulsifying of fats.

Emulsification means making into small droplets which may be more easily attacked by digestive enzymes. Small droplets have a larger surface area for attack.

The bile is coloured by pigments that are produced from the breakdown products of haemoglobin. These pigments are *biliverdin* (greenish in colour) and *bilirubin* (reddish in colour).

The bile juice is stored in the gall bladder and empties into the duodenum by way of the bile duct when required.

The pancreatic juice. This is a secretion produced in the pancreas. It contains enzymes for digesting carbohydrates, proteins and fats in an alkaline medium. The pancreatic juice consists of the following:

Water

Alkaline salts, e.g. Sodium bicarbonate which neutralizes the hydrochloric acid from the stomach.

Trypsin in the form of trypsinogen. This is an inactive form of trypsin. (Enterokinase, in the intestinal juices lower down the intestine, activates trypsinogen). Trypsin is a protease which changes peptones into amino-acids.

An amylase enzyme (amylopsin) which changes starches into malt sugar (maltose).

A lipase enzyme (steapsin) for breaking down fats to fatty acids and glycerol. Stimulation of the duodenal lining by stomach contents causes it to secrete hormones *secretin* and *pancreozymin*. These hormones (together with some nervous action) regulate the output of pancreatic juice.

(b) Digestion in the small intestine is brought about by the enzymes in the pancreatic juice and those in the *intestinal juice* (succus entericus). The intestinal juice is secreted by glands within the lining of the small intestine. Intestinal juice contains:

Water.

Salts—sodium carbonate and bicarbonate.

Protease enzyme (erepsin) which converts peptones into amino-acids.

Sucrase (invertase) which converts cane sugar into glucose + fructose.

Maltase which converts maltose into glucose + glucose.

Lactase which converts lactose into glucose + galactose.

(c) *Absorption* takes place along the length of the small intestine. The surface area for absorption is increased greatly by the presence of villi.

The end products of digestion are absorbed by the walls of the small intestine, and pass to the areas shown as follows:

Glucose and amino-acids—into the blood capillaries
of the villi

Fatty acids and glycerol—into the lacteals of the
villi where they re-
combine to form fats

The fat droplets in the lacteals are carried to the general lymphatic system (see p. 116). The lymphatics empty by way of the thoracic duct into the venous system. In this way fats enter the blood circulation.

THE LARGE INTESTINE

The ileum leads into a wider sac-like structure known as the *caecum*. A blind-ended *appendix* leaves the caecum. The appendix aids the process of plant digestion and is therefore quite long in herbivores. The large intestine ascends from the caecum on the right-hand side. It passes across the body just below the liver and descends on the left-hand side of the abdomen.

This wide sacculated tube is referred to as the *ascending colon*, the *transverse colon* and the *descending colon*. The continuation of the latter within the pelvic cavity is the *pelvic* or *sigmoid colon*. This becomes the *rectum* for the last 15 cm or so.

The rectum leaves the body by an opening known as the *anus*. The anus is guarded by an anal sphincter muscle. The large intestine has the same four layers in its wall as the small intestine, but its inside layer is not folded.

The functions of the large intestine are as follows:

(a) To prepare the *faeces* for removal from the body. The faeces are the solid "droppings" from the rectum. Their evacuation is known as *defaecation*.

(b) Water is absorbed from the contents of the large intestine; glucose and some salts are also absorbed.

(c) The rectum is lubricated by a *mucin* secretion produced in the large intestine. This allows the faeces to pass with ease.

(d) Some bacterial decomposition of plant material (e.g. cellulose) takes place in the large intestine. In order that the large intestine functions correctly and expels faeces by peristaltic action, an amount of undigested "roughage" must be present in the intestine.

The *faeces* are normally a brownish colour caused by the bile pigments, which are themselves breakdown products of haemoglobin. In liver disease, or when the bile duct is obstructed, the faeces become colourless or a pale straw colour.

The faeces are the rejected materials of our diet, they consist of cellulose, bacteria, mucin and water and food which has escaped digestion. The odour of the faeces comes from the bacterial decomposition of nitrogen compounds.

Biologically the faeces are not usually referred to as excretory products because they are not the by-products of our body chemistry but the unusable components of our food and other materials from the intestines.

Defaecation—the expulsion of faeces is partially a reflex action due to the presence of faeces in the rectum. They are squeezed into the rectum from the pelvic colon. About 100 g of faeces are passed daily depending upon food intake.

THE LIVER AND GALL-BLADDER

Liver-structure. The liver is similar to the salivary glands and pancreas in being an outgrowth of the alimentary canal (p. 66). It is situated beneath the diaphragm and is divided into left and right lobes, in fact about seven-eighths of the liver is on the right-hand side. This is the largest organ in the human body and it weighs about 1400 g. Embedded in the undersurface of the right lobe is a small sac called the *gall-bladder*. Blood from the stomach and intestine and the spleen all pass to the liver by way of the *hepatic portal vein*.

The liver also receives blood from the *hepatic artery*. Blood leaves the liver in the *hepatic vein*.

Stomach   carrying glucose

Intestines amino-acids–some fats → hepatic portal vein → liver → hepatic vein

oxygenated blood

Aorta ⟶ hepatic artery

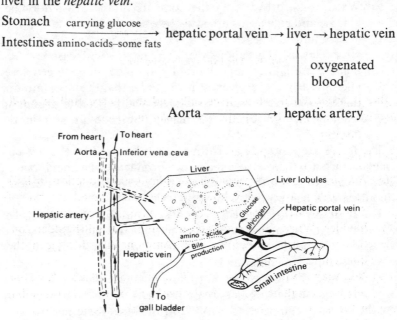

FIG. 38. The liver and blood-vessels.

The cells of the liver are arranged radially around a central intra-lobular tributary of the hepatic vein. These are sometimes described as *liver cords*. Between these rows of liver cells run blood *sinusoids* containing the blood coming from the intestines (hepatic portal blood) and the fresh blood coming from the aorta (hepatic artery blood). This blood runs through the liver and enters the central liver lobule veins.

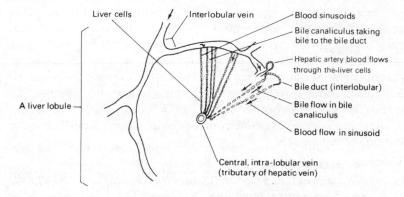

FIG. 39. The liver micro structure.

Bile is produced by some liver cells and this is secreted into *bile canaliculi* which run towards the *interlobular bile ducts*.

SOME FUNCTIONS OF THE LIVER

(a) *Bile production* and secretion is an important function of liver cells. Bile is a viscous yellowish-green fluid secreted at a rate of 500 to 1000 cm³ per day. This bile juice is stored in the gall-bladder. When fat enters the duodenum the gall-bladder contracts to push bile down the common bile duct into the duodenum.

Bile juice contains *pigments, bilirubin* (reddish) and *biliverdin* (greenish) both being produced when red blood cell haemoglobin is broken down. These are expelled with the faeces.

Bile juice contains *alkaline salts* which enable fats to be emulsified prior to digestion. The salts are *sodium glycocholate* and *taurocholate*. These salts are generally reabsorbed from the intestine and returned to the liver.

(b) *Storage* of glucose in the form of *glycogen*.

The storage of glucose as glycogen is regulated by a hormone. So

is the breakdown of glycogen. This can be seen summarized
below.

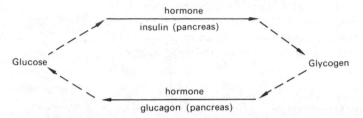

Fats, some vitamins (B_{12} anti-anaemic factor), iron and some
proteins are also stored in the liver.
(c) *Synthesis* of the important factors involved in the blood-clotting
sequence, prothrombin and fibrinogen.
(d) *Detoxification* of certain potentially poisonous materials, for
example, alcohol and aspirin.
(e) *The production* of the important anti-coagulant *heparin*.
(f) *Urea production* from the nitrogen containing portion of some
amino-acids.

THE PANCREAS

This is a glandular organ situated beneath the stomach, it is tongue-
shaped with the broadest edge on the right nearest the duodenum. It's
glandular secretions which enter the intestine do so through the pancreatic duct in common with the bile duct.
The pancreas is a dual purpose gland:
(a) Digestive secretion—pancreatic juice produced by most of the
gland (an exocrine secretion).
(b) Hormone secretion—*insulin* and *glucagon* produced by small
regions of the gland (an endocrine secretion) called *Islets of
Langerhans*.
The digestive juice which makes up pancreatic juice contains the
following:
Pancreatic amylase—splits starch to maltose.
Pancreatic protease—splits proteins into smaller units.

Pancreatic lipase—splits fats to glycerol and fatty acids.

Sodium bicarbonate—neutralizes stomach hydrochloric acid.

About 1200–1500 cm³ of pancreatic digestive juice is secreted daily. The secretion is produced by cells *(acini)* surrounding a central duct. Each group of *acini* or tubular structures is embedded within a background of connective tissue. The secretion passes into the duodenum by way of the pancreatic duct.

The hormone secretion contains:

Insulin—converts glucose to storage glycogen.

Glucagon—converts glycogen to glucose.

These hormones are secreted by the *Islets of Langerhans* which are islands of cells within the pancreas making up 1–2% of pancreatic tissue.

Hormones are secreted directly into the blood-stream, they are produced by *ductless* glands (endocrine glands).

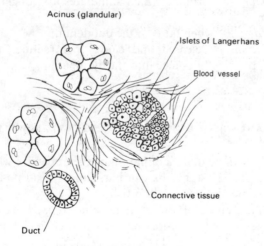

FIG. 40. Pancreas cells.

THE CONTROL OF DIGESTIVE SECRETIONS

The secretion of digestive juices is largely stimulated by the presence of food in the alimentary canal.

In the mouth the secretion of saliva is under nervous control.

In the stomach and duodenum the secretions are controlled by nerves and hormones.

Bile juice is caused to leave the gall-bladder by hormonal action.

In the intestines the secretion of intestinal juice is caused by the presence of food.

Gastric action. Nervous stimulation—vagus nerve.

Gastrin—a hormone released when food is in the stomach or duodenum, it is absorbed into the blood and stimulates the stomach glands to action.

Mechanical stimulation of the stomach lining.

Enterogastrone—a hormone released by the stomach and small intestine when fat is in the food, it is absorbed into the blood and inhibits stomach activity and delays its emptying.

Pancreatic action. Nervous stimulation—vagus nerve.

Secretin—a hormone which stimulates a weak pancreatic secretion.

Pancreozymin—a hormone which stimulates a stronger pancreatic secretion.

Gall-bladder action—when fat is in the duodenum.

Cholecystokinin is released and causes contraction of the gall-bladder, pouring bile-juice into the duodenum.

SOME DISEASES AND DISORDERS ASSOCIATED WITH THE DIGESTIVE SYSTEM

Appendicitis. An acute inflammation of the appendix which can be fatal if untreated. The appendix can burst and spread its infection to the abdominal membranes (peritonitis).

Constipation. Failure of the large intestine to expel its faeces. The causes of occasional constipation may be a change in diet or a decrease in fluid intake. (See function of roughage in the diet.)

Diarrhoea. This is the passage of semi-solid or fluid stools. It may be caused by some irritant in the food, a living organism or a chemical substance. In sensitive people anxiety and fear may cause diarrhoea. Disease of the lower digestive tract may be a cause of more persistant diarrhoea.

Dyspepsia (Indigestion). A general term covering upsets arising from the stomach with symptoms of pain, flatulence and nausea. "Heart-burn" being acid from the stomach being thrown up into the throat whilst regurgitating.

Flatulence ("wind"). Wind in the lower intestines may cause some abdominal discomfort. It may be taken in during swallowing or by fermentation in the lower intestine.

Gall-stones. Fragments of insoluble material deposited from bile juice. These "stones" may block up the common bile duct and thereby prevent the flow of bile juice into the intestines. Intolerance of fatty foods and jaundiced conditions may accompany this gall-bladder blockage.

Gastro-enteritis. An inflammation of stomach and intestine linings. Symptoms such as vomiting, abdominal pains and diarrhoea may result from a range of disorders of the stomach and intestine. Gastric 'flu and summer diarrhoea are virus-caused types.

Hepatitis. An inflammation of the liver tissues brought on by a virus or some toxin. Symptoms are similar to those of gall-stones, that is the jaundiced condition with intolerance of fatty foods, yellowish skin because of bile pigments in the blood, clay-coloured faeces and rust-coloured urine.

Mumps. A disorder caused by an air-borne virus. It causes a painful swelling of the parotid salivary glands. In grown men mumps sometimes causes sterility.

Thrush. A fungus infection of the mucous membranes, usually in the mouth. Characterized by white patches in the mouth.

Ulcers. A break in the body surface, the skin or mucous membrane which fails to heal. *Peptic ulcers* are ulcers caused by the stomach protein-digesting enzyme pepsin. This enzyme mixed with stomach acid becomes active and digests flesh. It is not normally able to do this because of the protecting film of mucous over the intestine and stomach linings. Gastric ulcers are in the stomach, duodenal ulcers are found in the duodenum.

CHAPTER 5

Respiration

RESPIRATION will be considered under two headings:

External respiration. This is the exchange of gases between the body cells and the external environment. The gases exchanged are *oxygen* and *carbon dioxide. Breathing* is the process of ventilating the lungs.

Internal respiration. This is the exchange of gases between cells and their fluid, internal environment. The gases involved are related to the internal chemical activity within cells. The oxidation of glucose within cells is called *tissue-respiration.* Oxygen is used, carbon dioxide is given out.

EXTERNAL RESPIRATION

(i) The Respiratory Organs

Air is taken into the nostrils and passes through the air-conducting passages before reaching the moist respiratory lung surfaces which are in close contact with the blood flowing within the lung capillaries.

NASAL CAVITIES

The nostrils open into the nasal cavities which are lined by ciliated epithelium. The movement of these cilia drives the mucus and dust particles from the throat area towards the nostrils. Air circulating through these cavities is warmed and cleared of particles. It is also moistened. The back of the nasal cavities opens into the pharynx through which both air and food pass.

The pharynx has been considered already under the study of the digestive system. As this passage carries both food and air some means of preventing food entering the respiratory passages is necessary. The posterior soft palate and the epiglottis over the larynx help to prevent choking.

THE LARYNX

The larynx is a modification of the air passage between the pharynx and the trachea (windpipe). This structure, which is commonly known as the "Adam's apple" or "voice box", has at its upper opening a leaf-like "lid" which closes over the opening to the windpipe when food is swallowed. This is called the *epiglottis*. The larynx is supported by hyaline cartilage. The larynx contains the vocal cords, which vibrate to produce the voice. Whispering does not involve the vocal cords.

The trachea is a short 10 cm tube running from the larynx to the upper margin of the chest where it divides into two branches. The trachea or windpipe is an elastic tube held open by C-shaped hoops of cartilage. It is about 2·5 cm in diameter.

The lining within the trachea secretes mucus which is kept in constant movement upwards towards the throat, by means of the cilia on the epithelial lining. Dust is trapped within the mucus of the windpipe and this prevents its entry into the lungs.

The thyroid gland lies across the upper end of the trachea just beneath the "Adam's apple" (larynx).

The bronchi are the two tubes at the base of the trachea which run into smaller air passages called the *bronchioles*. The bronchioles conduct air into the *air-sacs* (*alveoli*) of the lungs.

The bronchi and bronchioles are lined by a ciliated, mucus-secreting membrane resembling that located in the trachea.

The lungs are the respiratory surfaces where oxygen is taken up into the blood. No gas exchange takes place across the bronchioles which have thickish walls, they just conduct air to the lungs.

The left lung is divided into two lobes, an upper and a lower lobe.

The right lung is divided into three lobes.

Each lung is enclosed within a thin double membrane, the *pleura,* which enclose a thin film of fluid.

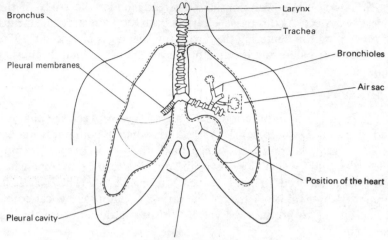

Bronchus

Pleural membranes

Larynx

Trachea

Bronchioles

Air sac

Position of the heart

Pleural cavity

Dome shaped diaphragm

FIG. 41. The respiratory organs (relaxed position).

(ii) Breathing

Air moves in and out of the lungs with a constant and rhythmical change in capacity of the thorax, produced by muscular action. The *mechanics of breathing* are as follows:

Inspiration—the intake of air to the lungs.

(a) The rib or intercostal muscles contract; the ribs and breast-bone move upwards and outwards.

   The capacity of the chest increases as the following takes place.

(b) The diaphragm moves downwards as it contracts (it is dome-shaped when relaxed).

The above two actions increase the capacity of the thorax and the elastic lungs expand to fill the increased thoracic cavity. Air is sucked into the lungs from the atmosphere in an effort to equalize the pressure.

Expiration—the expulsion of air from the lungs.

(a) The rib or intercostal muscles relax; the ribs and breast-bone move downwards and inwards.

   The capacity of the chest decreases as the following takes place.

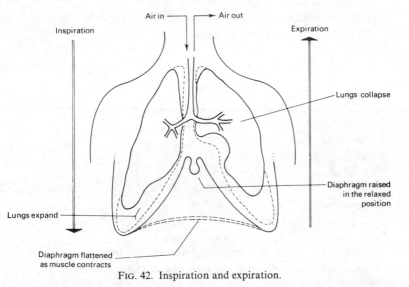

Fig. 42. Inspiration and expiration.

(b) The diaphragm moves upwards as it relaxes.
The capacity of the thorax is diminished and the elastic lung tissue contracts and air is forced out of the lungs into the atmosphere.

ARTIFICIAL EXTERNAL RESPIRATION

When the intercostal muscles and diaphragm are unable to ventilate the lungs, artificial breathing must be employed. *Mouth-to-mouth respiration* may be employed in an emergency.

The essentials of mouth to mouth resuscitation are as follows.

(a) An unconscious person may suffocate because the tongue can fall back across the pharynx and block the air passage. Pull the head back so that the jaw sticks in the air. Pull the lower jaw forward. The mouth will come open slightly and unblock the air passage by bringing the tongue foward away from the back of the throat.

(b) Put the lips around the mouth of the unconscious person, hold his nostrils closed, keep the jaw in the other hand. Blow air into the person's lungs about 12–15 times per minute until the chest is seen to move. In between each "blowing-in" allow expiration to take place naturally.

(c) An alternative is mouth-to-nose respiration. In this case the person's mouth must be kept closed.

(iii) The Capacity of the Lungs

Quiet breathing involves the movement in and out of *tidal air*. The *tidal volume* is about 500 cm³ per breath.

Man breathes in and out about 16 times per minute when at rest.

Forced inspiration allows a further 2500 cm³ more air to be inspired —this is called the *inspiratory reserve*.

Forced expiration allows a further 1000 cm³ more air to be expelled —this is called the *expiratory reserve*.

The lungs cannot be emptied because 1000–1500 cm³ air always remains, this is called the *residual volume*.

The *vital capacity* is the largest volume of air which can be expired after the deepest possible inspiration. This may be used as a measure of physical fitness. Only the very fit are able to show a vital capacity of 4000 cm³.

At rest a man breathes, for example, 16 times per minute. He takes in 500 cm³ per breath. His *respiratory-minute-volume* is therefore (16 × 500 cm³) 8000 cm³. The lungs are ventilated each minute by 8 litres of air.

This ventilation increases greatly in exercise, as much as 200 litres has been recorded.

(iv) The Composition of Respired Air

In quiet breathing 500 cm³ of air is taken in with each breath. This air contains

oxygen	21·00 %	volume
carbon dioxide	0·04 %	,,
nitrogen	79·00 %	,,

Expiration moves 500 cm³ of air out with each breath. This air contains

oxygen	16·00 %	volume
carbon dioxide	4·00 %	,,
nitrogen	79·00 %	,,

This expired air is saturated with water vapour.

Oxygen is taken up from the air in the air-sacs, and is passed into the blood. (By diffusion across the lung tissue into the blood capillaries surrounding the air-sacs.) The blood pigment haemoglobin picks up oxygen and as a compound *oxy-haemoglobin* releases this oxygen to the tissues needing it.

Carbon dioxide is passed out of the blood into the air-sacs.

(v) Control of Respiratory Movements

Quiet breathing movements are involuntary and take place automatically because of regular nervous impulses from the brain *respiratory centres,* in the medulla oblongata.

A most important factor in the regulation of the respiratory centre is the level of carbon dioxide in the blood. Increases in carbon-dioxide level cause an increase in respiratory movements.

Decreases in the carbon dioxide level in the blood bring about a decrease in respiratory activity.

INTERNAL RESPIRATION

The lung ventilation processes have brought oxygen into the blood. The red pigment within red blood cells picks up the oxygen and transports it to areas of the body in need of oxygen. Areas such as those growing, or carrying out some vital function, require oxygen.

Oxygen is needed for its part in the oxidation chemistry within the cell. Glucose, which is also carried in blood to the cells, is oxidized within cells and energy is produced. This cell oxidation process is usually summarized as below (tissue respiration equation):

$$C_6H_{12}O_6 \ + \ 6O_2 \ \longrightarrow \ 6CO_2 \ + \ 6H_2O \ + \ \text{Energy}$$

glucose   oxygen   carbon   water
                              dioxide

Tissue respiration is a complex chemical sequence when a carbohydrate is oxidized to produce energy. The chemistry involved in this is too complex for this level of study, but may be summarized in another way as below:

Stage 1. Glucose is oxidized to pyruvic acid

Stage 2. Pyruvic acid is put into a chemical sequence called *Kreb's cycle.*

This acid is oxidized and carbon dioxide and water are produced. The oxidation for most of this chemistry is by removal of hydrogen. *Dehydrogenase* enzymes are responsible for this.

Both these reaction chains give off energy.

Both these reaction sequences need oxygen.

If not enough oxygen is available (in extreme exercise) then after stage 1 the pyruvic acid becomes converted to *lactic acid.* It is this acid that causes the sensation of muscular fatigue.

Rest is necessary after this fatigue in order to "pay back" the "oxygen debt", thereby getting rid of the lactic acid in the muscles by oxidation.

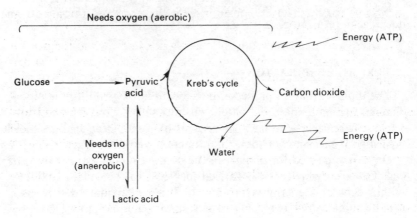

FIG. 43. Tissue respiration.

Energy is stored in a chemical called ATP (adenosine triphosphate). The energy from tissue respiration is stored in cells as this specially high energy phosphorus compound. When energy is needed from this store the compound is broken down to give up its stored high energy phosphorus. This energy is particularly needed for muscle activity.

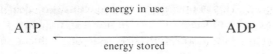

Adenosine diphosphate will take up and "store" energy in another phosphorus molecule and become adenosine triphosphate.

SOME DISEASES AND DISORDERS ASSOCIATED WITH THE RESPIRATORY SYSTEM

Asthma. A tendency to experience difficulty in breathing. As the person breaths out involuntary muscles around the bronchi contract, squeezing these tubes and making it difficult to expel air. There is a characteristic wheezing as he breathes out. Asthmatic attacks may be brought on by a variety of stimuli, emotional and physical.

Bronchitis (acute). An inflammation of the air passages in the lungs. Usually spread by a virus infection from the nose and throat, often followed by a bacterial infection.

(Chronic). An inflammation of the air passages in the lungs of a long-lasting nature (sometimes now called English Disease). Characteristic is the frequent cough such as comes with smoking cigarettes.

Common Cold. A virus infection of the mucous membranes of the nose and throat. It is often followed on by an infection by bacteria. The usual symptoms are sneezing, tickling throat, running eyes, a cough and perhaps a headache. It usually clears up in a day or two. One does not "catch colds" by being in draughts or wet.

Hay-fever. A seasonal allergy to the pollen of grasses, etc., found in some people. The mucous membranes in the nose are sensitive and react by an acute inflammation producing sneezing and watering of the eyes.

Influenza. An infectious virus disease of the respiratory system but has affects upon the whole body with fever, headache and weakness.

Laryngitis. This is an inflammation of the larynx. It may follow as a complication after a common cold. It may be caused by a spread of infection downwards from the nose and pharynx. A sore throat, a dry cough and hoarseness are symptoms.

Lung Cancer. This is one of the leading causes of death in modern society. The causes of the growth of tumours in the lung is a subject of active research. The association between lung cancer and tobacco smoking is strongly suggested.

Pleurisy. An inflammation of the pleural membranes around the lungs, caused by virus or bacteria.

Pneumonia. An inflammation of the lungs because of infection by viruses or bacteria. The infection is low down in the alveoli (air-sacs) and thus oxygen uptake is reduced.

Tuberculosis (pulmonary). This is a dangerous and infectious disease caused by a bacillus *(Mycobacterium tuberculosis)* which is usually inhaled. Inflammation of the lung tissue may leave scars after healing.

CHAPTER 6

Blood and Circulation

BLOOD is a connective tissue, fluid in nature and conveyed around the body in a system of closed tubes. The blood is kept on the move by a muscular pump, the heart. An adult man has about a gallon of blood (5–6 litres) making up one-twelfth (8 %) of the body weight of a 70-kg man.

The structures associated with the blood and its circulation are as follows:

(i) *The heart*—a muscular pump in constant activity.

(ii) *The blood vessels*—closed tubular vessels named arteries and veins. They complete a circuit by linking up as capillaries.

The functions of blood and the body fluids derived from the blood are as follows:

(i) *Respiratory function*—the transportation of oxygen and carbon dioxide.

(ii) *Nutritive function*—the transportation of food materials such as glucose, amino-acids and fats.

(iii) *Excretory function*—the carriage of excretory products away from tissues, such as urea and uric acid.

(iv) *Water supply to tissues*—fluid passes out from the blood to make up the tissue fluid which bathes the cells of tissues.

(v) *Protective and regulative function*—antibodies and hormones, etc., are transported in the blood. Some white cells are phagocytic and destroy bacteria.

(vi) *Temperature regulatory function*

The structures and functions just mentioned will now be studied in more detail, commencing with the blood itself.

93

THE BLOOD

Blood has solids suspended in fluid; *blood corpuscles* suspended in *plasma*. If blood is centrifuged in a tube (spun, rotated at high speed) the solids are thrown to the bottom of the tube because they are heavier, and the plasma lies above it. The blood can be separated into its two main parts in this way.

The yellowish plasma makes up 55 % of the volume of blood.

The packed mass of cells at the bottom of the tube make up about 45 % of the volume.

The measurement of the proportion of cells to plasma is done in a piece of laboratory apparatus called a *haematocrit*.

Blood has a pH of 7·3–7·4 (slightly alkaline) and a specific gravity of 1·055–1·065.

The Composition of the Blood

The *cells* or corpuscles of the blood are as follows:
    Red corpuscles (erythrocytes)
    White corpuscles (leucocytes)
    Platelets (thrombocytes)
The *plasma* contains the following:

Proteins (6·7 %)	serum albumin, serum globulin, fibrinogen.
Other proteins	some hormones, antibodies, enzymes.
Inorganic materials	salts of sodium, potassium, calcium, iron, magnesium, etc.
Organic materials (non-protein)	excretory substances such as urea; some glucose, amino-acids, fats, some hormones.
Respiratory gases	oxygen, carbon dioxide.

These components will now be looked at in a little more detail.

Red Blood Corpuscles (Erythrocytes)

The structure of the red cell is seen in Fig. 44. They have no nucleus. The diameter is about 7·2 microns (0·0072 mm) and the thickness is

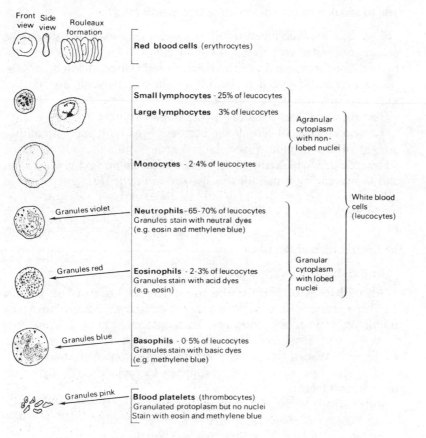

Fig. 44. Blood cells.

about 2·2 microns. The central area is very much thinner and gives the biconcave appearance typical of red cells. These biconcave discs tend to adhere together forming *rouleaux* when blood is split. A rouleaux resembles a pile of coins. Within the fatty-protein membrane of the red cell is a red pigment, *haemoglobin*. This oxygen-carrying pigment molecule contains four atoms of iron and is able to link up loosely with four molecules of oxygen. Every 100 cm³ of blood can carry up to 20 cm³ of oxygen in loose combination with haemoglobin.

THE ORIGINS, LIFE AND DEATH OF A RED BLOOD CELL

Red cells are manufactured in massive numbers daily within the medullary cavity bone marrow of skull, ribs and vertebrae bones. It is the red bone marrow that is involved in red cell production, not the yellow marrow. The cell begins life as a nucleated structure but this nucleus is lost very early in the development.

The average life of a red cell is four months.

Large quantities of blood cells are destroyed daily, something like 50 cm^3 of blood is destroyed daily. Throughout the body there are *phagocytic* cells which engulf old broken-up red blood cells. The spleen and liver are responsible for the disposal of the final fragments of red cells.

The haemoglobin is broken up in the liver to form the bile pigments bilirubin and biliverdin. These are the pigments which colour the faeces.

The iron from broken down haemoglobin is kept in the liver and used again for rebuilding haemoglobin.

An odd form of haemoglobin is sometimes produced in the red cells and associated with this are the cells which are abnormally fragile, they burst and lose their haemoglobin. This is a hereditary disease called *sickle-cell anaemia* and is found in parts of Africa. Having the disease seems to give one protection against malaria.

The formation of new red corpuscles to replace those dying off requires two factors:

> *An extrinsic factor*—vitamin B$_{12}$, a cobalt-containing compound called *cyanocobalamin*. Stored in the liver.
>
> *An intrinsic factor*—a protein produced by the stomach which seems to be necessary for the absorption of vitamin B$_{12}$ from the intestines.

These two factors combine to produce the *haemopoietic* factor (red-cell development factor). This is stored in the liver and passes from there to the bone marrow to influence red cell production.

Red-cell Count

The number of red corpuscles in 1 cm^3 is 4·5 million for women and

5·0 million for men. A piece of equipment called the *haemocytometer* is used for counting the numbers of blood cells in one cubic millimetre. The red cell numbers will vary according to circumstances, numbers are greater at high altitudes, in warmer atmospheres and during muscular exercises.

Red-cell Functions

Oxygen transport is possible because of the red pigment haemoglobin which enters into a loose combination with oxygen to form *oxy-haemoglobin*. The oxygen is released at areas of need.

Carbon-dioxide transport is carried in part by the red cells, as a direct combination between the gas and haemoglobin. The greater part of carbon dioxide is carried in the plasma.

White Blood Corpuscles (Leucocytes)

The structures of white cells are seen in Fig. 44. They have various shapes, have a nucleus but no coloured pigment. By staining a blood film *five* different types of white cell are distinguished.

(a) Those with *no granules* (agranular) in the cytoplasm.

    Lymphocytes and *Monocytes*

(b) Those *with granules* (granulocytes) in the cytoplasm.

    Neutrophils, Eosinophils and *Basophils*

THE ORIGINS AND LIFE OF A WHITE BLOOD CELL

The agranular white cells are produced in the bone marrow alongside the red cells.

The lymphocytes are also produced in lymphoid tissue such as the spleen.

Some lymphocytes have a short life; others may live as long as 6 months.

The granulocytes are produced in bone marrow and their life may be only a few days.

White-Cell count

The number of white corpuscles in 1 cm³ of blood is between 5000 and 10,000.

White cells are counted in a similar way to red cells, by using a haemocytometer.

The numbers of white cells normally increase 3 or 4 times in number during a bacterial infection. An abnormal increase in white cell numbers with no apparent cause can be a cancer-like *leukaemia*.

White-cell Functions

Lymphocytes increase greatly during an infection because they are involved in the production of protective *antibodies*. These substances are liberated from lymphocytes in response to the presence in the body of foreign proteins such as those of bacteria and viruses. These foreign proteins are named *antigens*. An antibody is a complex protein (globulin) which inactivates the antigen, rendering it harmless.

Antibodies are specific in their action. Antibodies against the measles virus have no protective action against chickenpox virus.

Monocytes, unlike the lymphocytes, are capable of *amoeboid movement*. These cells engulf foreign substances within the body fluids by a process described as *phagocytosis* ("cell eating"). These cells can squeeze out of blood vessels and enter the tissue fluid in order to protect the tissues from invading bacteria. Many of the white cells are killed by the poisons put out by the invading bacteria. The dead white cells accumulate as a *pus*, and a *boil* or *abscess* may form.

Neutrophils are phagocytes which destroy foreign substances and bacteria by engulfing them and digesting them. It is the granules within the cytoplasm which contain the digestive enzymes.

Pus will contain many dead and living neutrophils.

Eosinophils produce an enzyme which can dissolve away blood clots. They also have *anti-histamine* action. (*Histamine* is released in tissues when they are injured and produce the inflammation symptoms. Allergies such as hay-fever cause histamine to be released in some tissues.)

These cells are not phagocytic and are not very mobile.

Basophils are not phagocytic and not very mobile in the way that the amoeboid type monocytes are.

These cells are not found in very large numbers in normal blood.

The functions of these cells are not altogether clear.

They are known to produce *heparin* (blood anti-coagulant) and *histamine* (the inflammation "causing" agent).

Blood Platelets (Thrombocytes)

The structure of the blood platelet is seen in Fig. 44.

They are very small (about one-quarter the diameter of a red cell); they have no nucleus and their cytoplasm contains granules.

The origin of platelets. They are made from large cells in the bone marrow. They are considered to be fragments broken off from these giant cells.

The numbers of platelets per cm^3 is about 250,000.

THE FUNCTION OF PLATELETS—BLOOD CLOTTING

Injury to the tissues with loss of blood is prevented by the action of thrombocytes. At a point of injury, a wound, blood platelets release a substance *serotonin* (a hormone) which causes blood capillaries to constrict, to reduce blood flow. Loss of blood from the wound is prevented by a plug of platelets forming. The formation of this plug or clot is brought about by the action of an enzyme *thromboplastin* (thrombokinase) released from damaged platelets when a wound is received.

THE FORMATION OF THE BLOOD CLOT

Blood plasma has been seen to contain a soluble protein *fibrinogen*. A clot is formed when this soluble protein is converted to an insoluble *fibrin*. Clotting involves blood platelets not red cells. Red cells are concerned in a process called *agglutination*. The sticky fibrin threads fill up the open wound as a jelly-like clot which hardens and contracts as exposed to air. The hardened, complete clot may take 2 hours to form. A clear fluid, *serum,* is squeezed out of the clot as it hardens.

A summary of the *clotting mechanism* is given below:

(a) *Thromboplastin* is released from damaged platelets and tissue.

(b) Conversion of a plasma protein *prothrombin* to the active enzyme *thrombin*. This takes place only when calcium ions and thromboplastin are present.

(c) This thrombin enzyme produced will break down the soluble *fibrinogen* to form the insoluble sticky threads of fibrin.

(a) Damaged platelets

  Damaged tissues $\longrightarrow$ Thromboplastin

(b) Prothrombin + Calcium ions + Thromboplastin → Thrombin
    (inactive enzyme)   (active enzyme)

(c) Fibrinogen + Thrombin ⟶ Fibrin
    (soluble in plasma)   (insoluble threads)

A blood clot does not form in the circulation under normal circumstances because there is not enough free thromboplastin present to start the clotting action. A safeguard against any clot forming in the normal circulatory system is the presence of *anti-thrombin* in the blood.

Thrombosis is a medical condition when a blood clot (a thrombus) is formed in the circulatory system. If this clot is in the heart blood vessels (the coronary vessels) then a serious heart "attack" may result. The clot is formed because of some injury or change in the nature of a blood-vessel wall.

FACTORS SPEEDING UP CLOT FORMATION

(a) Injury to the tissues or to the wall of a blood vessel.

(b) The addition of thrombin to the blood.

(c) The presence of calcium ions.

(d) An increase in temperature.

(e) Contact with foreign matter, gauzes, bandages. Various chemicals (styptics).

FACTORS PREVENTING CLOT FORMATION (anti-coagulants)

(a) Minimum tissue damage, smooth-lined blood vessels.

(b) Grease or oil on surfaces.

(c) Removal of calcium ions by the addition of sodium hydrogen citrate (as for collecting blood after blood donation).

(d) A decrease in temperature (transfusion blood stored at 4°C).

(e) *Heparin* is an anti-coagulant found in the liver, lung, muscle and other tissues. This substance prevents the conversion of fibrinogen to fibrin.

BLEEDING TIME AND CLOTTING TIME

Bleeding time is measured by pricking the ear lobe or base of the finger-nail and noting how long the bleeding continues. As a bead of blood is formed it is removed by a filter paper until such time, 2–6 minutes, that no more blood appears. It stops because a platelet plug forms and blood vessels close up in the area.

Clotting time on the other hand is the time taken for a clot to form. Blood is put into a grease lined tube and rocked in body warm water until the blood stops flowing. It takes less than 10 minutes to show the first signs of a clot.

Haemophilia

This is a disease where a person's blood takes an abnormally long time to clot. Normally the blood clots in, say, 6–7 minutes; in haemophilia the blood may remain fluid for over an hour. A small injury can cause great loss of blood and an internal injury can be particularly dangerous as the bleeding will not be obvious.

Haemophilia is a *sex-linked recessive character* (see genetics later); it occurs almost always in males and is transmitted only by females.

In haemophilia one factor in the blood-clotting sequence is missing, haemophiliacs are unable to manufacture it.

Frequent blood transfusions with healthy blood or plasma are needed to supply the missing factor.

The Plasma

Plasma is a sticky, straw-coloured fluid which is the vehicle of all the raw materials used in the body.

Plasma is composed of:

water	91%,
proteins	7%,
salts	0·9%,

other substances such as glucose, etc.

Plasma is slightly alkaline with a pH of 7·4.

Plasma proteins have very important functions:

Albumin is present in largish quantities (4·5 g/100 cm³ plasma). It is produced in the liver and has a smaller molecular size than the other protein, globulin. Low amounts of albumin in the plasma (as in liver disease) cause changes in osmotic pressure and water leaks from the blood into the tissues causing the puffy skin of *oedema*.

Globulin is a larger molecule than albumin and is present in blood in smaller quantities (2–7 g/100 cm³ of plasma). This is also produced in the liver, except one type of globulin, called gamma globulin, which is produced in lymphoid tissue of the spleen and other areas. Gamma globulin extract from the blood contains the antibodies to infectious disease.

Fibrinogen is a very large molecule again formed in the liver and present in the plasma but in much smaller quantities (0·25 g/100 cm³ of plasma). This protein has already been studied as the soluble material which may be converted into the fibres that make up the blood clot.

It is important to notice that these large protein molecules are too big to pass across the walls of blood capillaries and that if they do so something is wrong. Proteins in the urine are symptomatic of some disorder as will be mentioned in the chapter on excretion.

Blood Groups and Blood Transfusions

Blood groups date back to 1900 when the Austrian *Karl Landsteiner* discovered the blood groupings referred to as the A.B.O. system. He grouped blood as either A, B, AB or O. Up until this date any person suffering a blood loss took the risk of being given a blood transfusion which was likely to kill him or give him great pain.

The reason for the dangerous results of earlier transfusions was because incompatible bloods *agglutinate,* that is the red cells all stick together to form clumps. These clumps block up small blood

capillaries. Eventually these cells burst and release their haemoglobin.
The ABO system may be summarized as below:

(a) Red cell membranes have a different chemical structure from
person to person. They carry a substance called an *agglutinogen*.
Cells can be A or B type with respect to this substance. If the cell
has both it is called AB type, if it has neither it is called group O.
These A, B, AB or O blood groups are inherited with the A and
B type being dominant, the O group recessive.

(b) Plasma contains different protein structures (globulins) which
will cause red cells to clump together, if the wrong cells are put
into the plasma. These protein structures are called *agglutinins*.
These agglutinins are anti-A or anti-B, which means they will
clump together either A- or B-type cells. Plasma contains anti-A,
anti-B, both or neither.

The Rhesus factor is another important blood group. This group was
discovered in 1940 in experiments with Rhesus monkeys. A series of
new agglutinogens were found on the red cells, the most important
being the D factor. People whose blood has the D factor are named
Rhesus positive (Rh$^+$); they make up 85% of the population. People
without this factor make up 15% of the population and are called
Rhesus negative (Rh$^-$). In transfusion Rh$^+$ patients can receive Rh$^-$
blood but the other way round creates problems. *The rhesus negative
man* will become sensitized to rhesus positive blood and will produce
anti-D on the first transfusion. A second transfusion must be of the
correct Rh$^-$ and correct ABO group otherwise agglutination will
occur.

The rhesus negative woman should never be given rhesus positive
blood because she will then become sensitized to rhesus positive blood
as a result of the anti-D agglutinin in the plasma. This can be dangerous
if the woman becomes pregnant by a rhesus positive man producing
a rhesus positive child growing within her. Her anti-D antibodies may
filter across the placenta and damage the child's red blood cells.

A blood transfusion becomes necessary if a person loses more than a
litre of blood. The blood to be transfused must be compatible, that is,
not likely to agglutinate the blood of the recipient. It is best to give a
blood which is the same as that of the patient; if this is not possible
then group O may be given because its cells carry no agglutinogen.

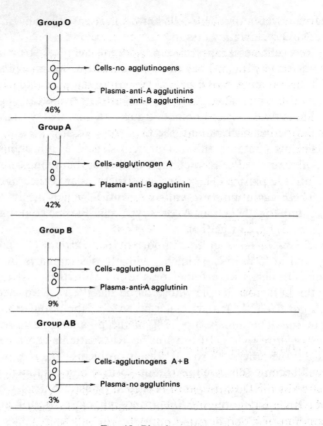

FIG. 45. Blood groups.

A blood may be transfused as shown below:

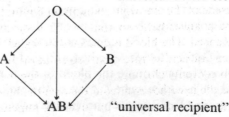

Before a transfusion, blood must be grouped, this may be done as summarized below:

(a) A serum containing known agglutinins is needed.
   A serum with anti-A and a serum with anti-B.

(b) The unknown blood is added to a tube of anti-A and to a tube of anti-B serum. The glass tubes are rocked from side to side for a short time.

Agglutination may or may not take place as shown in the table of results below:

	Anti-A serum	Anti-B serum
Unknown corpuscles		
O	—	—
A	+	—
B	—	+
AB	+	+

+ = blood agglutinates.

AT THE BLOOD TRANSFUSION CENTRE

The National Blood Transfusion Service is an organization co-ordinated by the Department of Health of the day, by whatever name. Blood is donated by volunteers who attend not more than twice a year. The amount of blood donated on each visit will be 440 cm³ (1 pint). Donors must not have any history of disease which includes the following: malaria, hepatitis (jaundiced), venereal disease, or any blood infections.

Donating blood involves blood being quickly drawn from a vein on the inside of the elbow, it takes about 5 minutes. A doctor gives a local injection anaesthetic so that no pain or unpleasant sensations are experienced. The blood is collected in sealed aseptic bottles containing 120 cm³ of acid citrate-dextrose solution for 440 cm³ of blood. The citrate prevents clotting, the dextrose sugar feeds the red cells. When the needle is withdrawn from the arm the blood stops flowing in a few minutes and a dressing is put over the puncture. Blood left over in the needle and in the tubing is put into a smaller pilot bottle attached to the side of the main collecting bottle. This pilot blood does not have any anti-coagulant added to it. The clot and serum separate. This is used for typing the blood as to ABO and Rh factors. The serum is used to test for venereal disease (syphilis). These groupings may be done by an automatic machine.

After the blood has been grouped and labelled it is stored at $+4^{\circ}C$. It will keep for 3 weeks at this temperature. Blood not used within this time has the plasma separated from the cells. The plasma is dried to a powder and may be used in an emergency when added to distilled water.

THE CIRCULATION OF THE BLOOD

The blood circulates round the body in closed tubes called blood-vessels. The fluid is kept on the move by a muscular pump called the heart.

Until 1628 it was believed that blood was made from the food we ate which passed from the intestines to the liver where blood was made. In 1628 William Harvey published work that he had done to demonstrate the continuous circulation of blood. Before his time it was thought that blood just moved back and forth in the vessels.

It was not until 1661 when Malpighi demonstrated capillaries in the tissues that the complete picture was given, because Harvey had no knowledge how veins and arteries linked up through the tissues. He suggested pores in the tissues allowed blood to pass through.

The circulatory system will be studied in two parts—the heart and its activity, and the blood vessels.

The Heart

The heart is rather like an inverted cone, made of a special (cardiac) muscle. The apex of this cone points forward, and to the left just above the diaphragm on the left hand side of the thorax, extending to just below the left nipple.

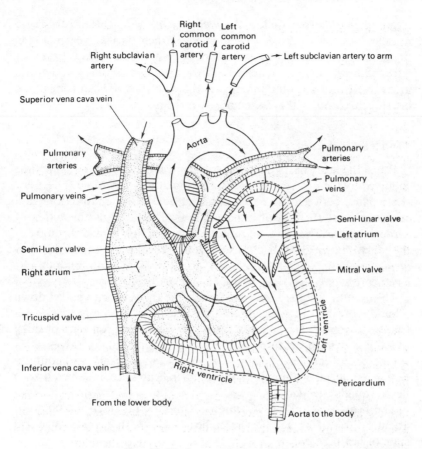

FIG. 46. The heart and great vessels.

The heart has four chambers, *two atria* (auricles) side by side and *two ventricles* side by side. The ventricles are enclosed within thick muscular walls. The atria are thin-walled and situated above the ventricles. The atria are connected with the ventricles by way of valves as seen in the diagram. The right valves are called the *tricuspid valves,* the left hand side valves are called the *mitral valves.*

The inside of the heart chambers are lined by a delicate membrane called the *endocardium.* Inside the ventricles the wall is extended into finger-like muscular papillae, the *papillary muscles*. The ends of these muscles have cords, the *chordae tendinae*, attached to the ventricle side of the valves separating the atria from the ventricles. The valves prevent a back flow of blood into the atria. Valves in the pulmonary and aortic arches have the same function, preventing back flow. The outside of the heart is enveloped within a sac known as the pericardium.

The Heart in Action

When the heart beats the two atria contract at nearly the same time followed by the two ventricles contracting at the same time. There is a short pause between these two contractions. These contractions and relaxations are rhythmical in nature and are caused by the action of cardiac muscle. The waves of muscular contraction spread throughout the much branched cardiac muscle from a focal point called the *sinuatrial node* in the wall of the right atrium. It is this S.A. node which acts as a heart *pacemaker*, originating each heart beat.

The excitation spreading from the atria to the ventricles passes down a bundle of fibres between the two ventricles called the *bundle of His* (the atrio-ventricular bundle). The two ventricles contract pushing blood out of the heart through the aortic and pulmonary valves.

One heart beat (a cardiac cycle of relaxation—*diastole* and contraction—*systole*) takes about 0·8 second. Each beat is accompanied by two heart sounds, *lub-dup*.

The heart beats about 75 times per minute (at rest) pumping out about 5 litres of blood. The heart beat varies with age, sex, size and state of health.

The cardiac cycle is a sequence of events during one heart beat. These events may be summarized as below:

(a) *Atria*—relaxation. Blood enters *right atrium* from the body through the vena cavae. Blood enters *left atrium* from the lungs through the pulmonary veins. Both atria fill up and the blood flow into them ceases.

Atria—contraction. Blood is pushed through the atrio-ventricular valves into the still relaxed ventricles. The ventricles become full and pressure forces the a.v. valves to close preventing a back flow of blood into the atria.

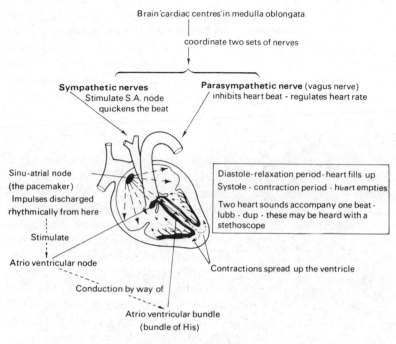

Brain 'cardiac centres' in medulla oblongata

coordinate two sets of nerves

Sympathetic nerves
Stimulate S.A. node
quickens the beat

Parasympathetic nerve (vagus nerve)
inhibits heart beat - regulates heart rate

Sinu-atrial node
(the pacemaker)
Impulses discharged
rhythmically from here

Stimulate

Atrio ventricular node

Diastole-relaxation period-heart fills up
Systole - contraction period - heart empties

Two heart sounds accompany one beat -
lubb - dup - these may be heard with a
stethoscope

Contractions spread up the ventricle

Conduction by way of

Atrio ventricular bundle
(bundle of His)

FIG. 47. The heart beat.

(b) *Ventricles*—relaxation. Blood received from the contracting atria.

Ventricles—contraction. Blood pushed against the semi-lunar valves at the base of the aorta and pulmonary arteries. Blood leaves the heart.

During this cardiac cycle *deoxygenated* blood is brought into the heart and pumped to the lungs for a new supply of oxygen and to get rid of carbon dioxide.

The *oxygenated* blood returning to the heart from the lungs is pumped around the body to supply the tissues with oxygen and to pick up carbon dioxide.

E.C.G. (Electrocardiography). Recording electrical changes in the heart muscle is a valuable way of investigating various heart abnormalities. These electrical changes can be amplified and recorded on a

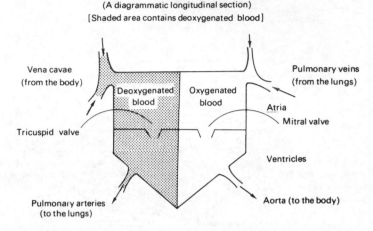

FIG. 48. Diagram to illustrate blood flow through the heart (cardiac cycle).

cathode-ray oscilloscope or by a pen recorder. The figure produced is an *electrocardiogram.*

Heart transplantation as a technique has much experimental history. The transplantation of one animal organ to another animal has been well practised. The problem of rejection by the host tissues is a major difficulty but when overcome in experiments with dogs it is seen that the heart can maintain an adequate blood circulation.

The transplanted heart continues to beat despite the cut nerve supply, and its regulatory system seems to function because it increases its activity in exercise.

The Circulation and the Blood Vessels

Blood leaves the heart in closed tubular vessels, circulates around the body and returns to the heart *(systemic circulation)*. Blood leaves the heart, passes into the lungs and returns to the heart *(pulmonary circulation)*.

These two circulations are shown in Fig. 49.

(a) *Systemic blood* is pumped from the left ventricle, as seen in Fig. 49. The blood leaving this ventricle is at a fairly high pressure

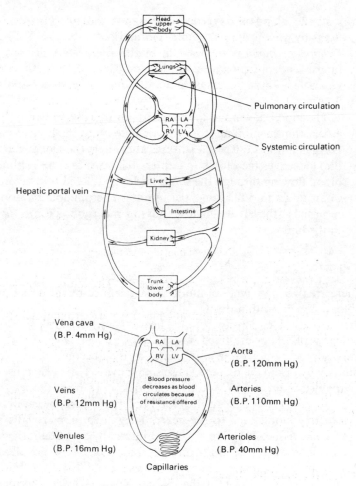

FIG. 49. The circulatory system.

because the blood must overcome the resistance of the vessels through which it flows. The finer the vessels the more difficult it is to force blood through them. After the blood has passed through the tissue capillaries the pressure falls and the veins convey the blood back to the right atrium of the heart. The systemic blood is oxygenated, it gains the oxygen from the lungs by way of the pulmonary circulation.

(b) *Pulmonary blood* is pumped from the right ventricle into the lungs. The pumping pressure is less than that of the other ventricle because the lungs do not offer much resistance to blood flow.

The blood flowing in the lung capillaries is separated only by a very thin membrane from the air taken into the lung air sacs. Oxygen diffuses out of the air sacs and enters the blood. Carbon dioxide leaves the blood and enters the air-sacs. The volume of blood flowing through the lungs is closely related to the volume of air taken into the lungs per minute. The pulmonary blood is returned to the left atrium after having undergone a *gas exchange* in the lungs.

The Blood Vessels

There are two main types of blood vessel, defined by the direction of the blood flowing within them.

Arteries carry blood away from the heart. They have thickish elastic, muscular walls. They are lined internally by an *endothelium*. Arteries gradually become narrower in diameter as they become closer to the capillary network in the tissues. These smaller vessels are also muscular and capable of contraction to restrict blood flow. They are the *arterioles* which offer some resistance to the blood before it enters the capillaries.

Veins carry blood back to the heart. They have thinner walls with less muscular tissue. They are also lined internally by an endothelium. The veins have valves within to prevent blood backflow. Small veins leading from the capillaries are called *venules*.

Capillaries pass in amongst the tissues. They have semi-permeable walls which allow the passage of small molecules from the blood to pass into the *tissue fluid* which bathes the tissues. From the tissue fluid small molecules then pass into the cells. These movements are diffusion processes.

The pulse is the pulsation felt beneath the skin at various points in the body where an artery comes close to the skin surface. The pulse corresponds to the heartbeat.

The pulse rate may vary within the following ranges:

60–80 per minute	resting adults
80–100 per minute	children 6–10 years
100–120 per minute	infants

The resting rate of the pulse is often less important than changes in the pulse rate after exercise.

Blood pressure (B.P.) is the pressure at which the heart pumps blood into the aorta and other main arteries. This B.P. can be 120 mm of mercury in the systolic pressure and 80 mm of mercury in the diastolic pressure (see Fig. 49).

The blood pressure is normally measured at the artery in the arm (brachial artery) where the pressure of blood is similar to that just leaving the heart (i.e. 120 mm Hg).

The instrument used for measuring blood pressure is called the *sphygnomanometer*.

Lowering of blood pressure can deprive the brain of sufficient blood and *fainting* may result.

Increase in blood pressure can overwork the heart and blood vessels. *Hypertension* or high blood pressure is fairly common in modern man, the causes of it being unclear, except where there is kidney disease or other known disorder.

SOME DISEASES AND DISORDERS OF THE CIRCULATORY SYSTEM

Anaemia. A collective name for a shortage of the oxygen-carrying pigment haemoglobin, found in red blood corpuscles. There is a shortage of oxygen for tissues throughout the body. The causes of anaemia are various, including deficiency of iron or excessive destruction of red cells as happens when red cells are destroyed by the malarial parasite.

Angina pectoris. Severe pain because of shortage of oxygen supply to the heart muscle through the coronary arteries. These arteries lose their elasticity in some people and so they are less able to expand to allow extra blood when required in exercise. The pain may be alleviated by sucking a tablet of glyceryl trinitrate which expands the coronary arteries.

Cardiac arrest. If blood circulation ceases then unconsciousness may follow in 10 seconds and death within a matter of minutes.

Coronary thrombosis. The blockage of a coronary artery or one of its branches by a blood clot. The heart muscle is thereby deprived of blood and dies. Pain is similar to that in angina but it persists even upon resting. This disorder kills more people than all other diseases, it may be one of the "stress" diseases of modern society.

Faint. A reflex action resulting from overactivity of the vagus nerves because of emotional or other reasons. This causes sudden unconsciousness because of the lowered blood pressure in the arterial supply to the brain. Blood tends to pool in the abdomen where the blood vessels relax and greater blood flow is permitted.

Haemophilia. A rare disease where prolonged bleeding is symptomatic. One of the factors in the blood-clotting sequence is absent and so injury, however small, may cause prolonged bleeding. This is a sex-linked recessive character, linked so that men only are haemophiliacs.

Hypertension (high blood pressure). This may be the symptom of some disease, i.e. kidney inflammation or it may be a disorder on its own. It is certainly not a symptom to be ignored as it increases the likelihood of suffering from some other more chronic ailment.

Leukaemia. This is a cancer-like disease of white blood cells. The white cells increase in number but do not mature to the state where they are able to protect the body against invading disease. Normal mature white cells are eventually replaced by immature useless cells.

Palpitation. The awareness of one's heart-beat.

Varicose veins. The abnormal swelling of veins, particularly in the legs. The valves in the veins which normally ensure a one-way flow of the blood become defective. There is a back-pressure of blood and the feet often swell. Skin ulcers can form.

CHAPTER 7

The Lymphatic System

TISSUE FLUID

The blood does not come into direct contact with the cells. Cells are bathed in a *tissue fluid* which is a "go-between" the cells and the blood.

Tissue fluid is formed from blood plasma which leaves the capillaries as they pass through the tissues. Only the plasma proteins do not leave the blood-vessels. They remain behind in the blood-vessels and help to draw the tissue fluid back into the blood. There is a two-way flow

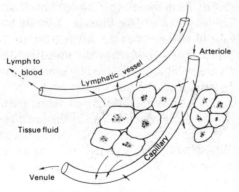

FIG. 50. Tissue fluid and lymph.

between the blood plasma and the tissue fluid as summarized in Fig. 50. Some of the tissue fluid is taken up in blind ended tubules belonging to the lymphatic system. The fluid within these *lymphatic vessels* is called

115

lymph and it closely resembles plasma but contains less protein. If an excess of tissue fluid is not drained away then it will accumulate to produce *oedema* (dropsy).

Lymph

This is the tissue fluid which is being returned to the blood circulation by way of a system of lymphatic vessels. The lymph passes through filtering units called the *lymph nodes*. "Filtered" in these nodes (or lymph glands) are "foreign materials" and bacteria taken up from the tissue fluid.

The lymph glands are mainly composed of lymphoid tissue. This tissue is also found in the *tonsils, adenoids, spleen* and as *Peyer's patches* in the lower part of the small intestine. The lymph nodes may be thought of as the safety barriers to prevent infective or dangerous materials from entering the blood from the tissues. The lymph nodes also produce *lymphocytes*.

The movement of lymph through the vessels depends upon the general activity and compression produced by muscles of the body. Back flow of lymph is prevented by valves within the vessels.

The lymph flows in the direction of the head and returns to the blood at shoulder level entering the veins, the left and right subclavians.

Fat absorbed from the intestines is passed into the lymph of the lacteals and in this way enters the general circulation.

SOME DISEASES AND DISORDERS OF THE LYMPHATIC SYSTEM

Adenoids. Lymph gland tissue above the soft palate and at the back of the nose. They are sometimes removed with the tonsils in recurrent sore-throat cases.

Elephantiasis. A disease caused by minute nematodes (thread worms) which invade the lymphatic system. The mass of worms block up the lymphatic vessels preventing the backflow of lymph.

Hodgkin's Disease. A disease characterized by enlargement of lymphoid tissue at the lymph nodes. Pain is caused as the nodes increase

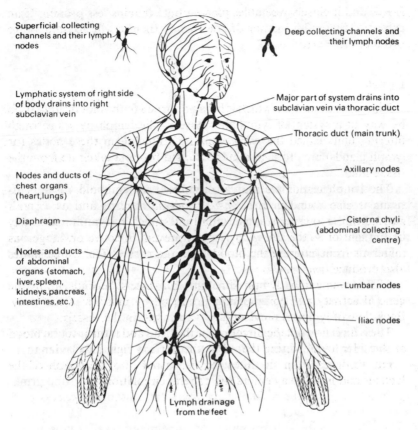

Superficial collecting channels and their lymph nodes

Deep collecting channels and their lymph nodes

Lymphatic system of right side of body drains into right subclavian vein

Major part of system drains into subclavian vein via thoracic duct

Thoracic duct (main trunk)

Nodes and ducts of chest organs (heart, lungs)

Axillary nodes

Diaphragm

Cisterna chyli (abdominal collecting centre)

Nodes and ducts of abdominal organs (stomach, liver, spleen, kidneys, pancreas, intestines, etc.)

Lumbar nodes

Iliac nodes

Lymph drainage from the feet

FIG. 51. The lymphatic system.

in size and press upon nerves. A chance infection may cause death because the diseased lymph nodes offer no protection.

Oedema. An excess of tissue fluid localized around some disturbed part of the body or throughout the body (dropsy). This tissue fluid is drawn from the blood.

Quinsey. An abscess around the tonsils. It may follow on from a sore throat.

Tonsillitis. A bacterial infection of the tonsils.

CHAPTER 8

Excretion

EXCRETION is the elimination from the body of potentially dangerous chemical substances that are the by-products of normal or abnormal body chemistry. Water, some salts, and other substances that are present in excess quantities are also excreted.

The main excretory substances and the major excretory organs are listed below:

The Kidneys   water, nitrogen compounds, salts, acids.
The Lungs   water and carbon dioxide.
The Skin   water and salts.
The Liver   bile pigments, salts and various toxic substances.

All the organs mentioned also have important functions other than excretion; these are studied elsewhere.

The excretory products listed have their origins in the food and drink taken in. They are the results of metabolic activity or are excesses of such as salts and water.

The chief metabolic by-products of some major food groups are as follows:

Proteins   urea, uric acid, creatinine, ammonia (nitrogen compounds).
Carbohydrates   carbon dioxide and water.
Fats   carbon dioxide and water.

The bile-pigments are formed in the liver from the break-down of old haemoglobin from old red blood cells.

Truly speaking *faeces* are not excretory products (except their colouring from bile-pigments) because they represent undigested food, never used by the body, on its way through the gut. In this section the

118

structure and function of the kidneys and associated parts will be studied.

THE URINARY ORGANS

The nitrogen compounds, salts and water mentioned previously are excreted in the urine which is produced in the *kidneys*. Urine passes from the kidneys by way of the *ureters* to the *bladder*. The exit from the bladder is called the *urethra*.

The Kidneys

Each kidney is situated at the back of the abdomen embedded in firm fat on either side of the backbone. The position of the liver forces the right kidney to be lower than the left kidney. The concave inner surfaces *(a hilum)* receive the renal arteries and the renal veins and from here the urine passes out from the *renal pelvis* and thence to the ureter.

Each kidney weighs more than 113 g. The ureters are about 30 cm in length and pass along the back of the abdomen to the bladder. The urine is propelled down the ureters by peristaltic action, just as food is moved along the small intestine.

The kidney in section shows the following main areas:

The pelvis is an open collecting area where the ureter leaves the kidney.

The pyramids are groups of collecting tubules emptying into the pelvis region by way of eight to twelve projecting openings. This inner area of the kidney, the *medulla* is paler in colour when seen in cross section.

The cortex is the outer part of the kidney and is darker in colour; it contains about 1 million minute twisted tubules called *nephrons*. It is these tubules that produce the urine from the blood flowing through the kidney. The whole kidney is enveloped in a fibrous tissue *capsule*.

THE FORMATION OF URINE

The minute structures shown in Fig. 53 (p. 121) are responsible

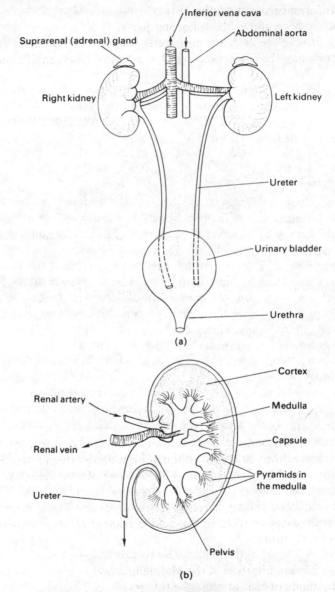

FIG. 52. (a) The urinary organs. (b) The kidney in section.

for the formation of the urine.

The nephron shows the following parts:

The *Malpighian body* (or corpuscle) made up of *Bowman's capsule* enclosing the *Glomerulus* which is a tuft of blood capillaries. The

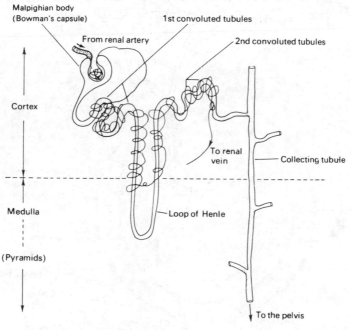

FIG. 53. A nephron.

Malpighian body or corpuscle was first described in 1666 by an Italian named *Malpighi* although he was not able to describe its function. Many years later (1942) an Englishman called *Bowman* described this structure as a plasma filtering unit.

The Tubules are very convoluted (twisted) but show three main areas: a first convoluted tubule, the *Loop of Henle,* a second convoluted tubule.

Urine is formed in the nephron by two processes:

   (a) plasma filtration in the Malpighian body,

   (b) fluid concentration in the tubules.

(a) *Filtration.* Blood enters the glomerulus through a wider arteriole and because the blood is then forced into much narrower capillaries a high pressure of blood acts against the walls of the blood capillaries. This high pressure forces small molecular substances across the walls of the capillaries and the walls of Bowman's capsule. These substances enter the kidney tubules.

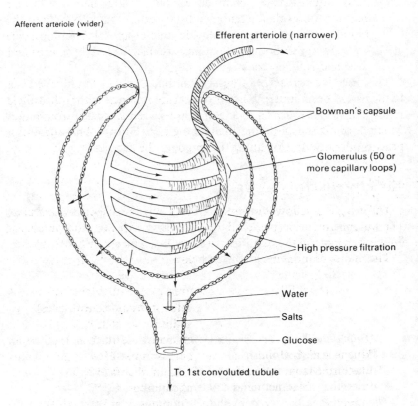

FIG. 54. Filtration in the Malpighian body.

The fluid passing across from the plasma into the tubules is not urine. Many of the materials that have passed over are taken back into the blood as they move down the tubules.

Each day about 170 litres of this filtrate is produced but the daily output of urine is nowhere near this, 1 to 1·5 litres of urine is usual. Most of this filtrate is reabsorbed in the tubules and taken back into the blood.

(b) *Fluid concentration* in the tubules returns about 98 % of the water filtered, back to the blood. Most of this is taken up in the first convoluted tubule, as much as 80%. Together with water reabsorption other essential substances are taken back into the blood: glucose, sodium chloride and other salts, amino-acids and vitamin C. Some potassium is reabsorbed. A little urea and uric acid is taken back.

This selective reabsorption returns all but 1 cm^3 of the filtrate back to the blood, per minute. Waste products are retained within the tubule to form urine. As much as 30 g of urea are excreted per day from blood which normally has no more than 30 mg per 100 cm^3. This suggests a process of concentration along the length of the tubules.

THE COMPOSITION OF URINE

The urine is a yellow fluid coloured by a substance called *urochrome*. The amount of urine passed daily will vary with the daily intake of fluid but in general the quantity passed is 1–1·5 litres.

The main components are listed below:

96% water		
4% solids	2% urea	
	2% salts (inorganic substances)	
	sodium	0·35%
	chlorides	0·6%
(Mucus will be found in	potassium	0·15%
the urine from the walls	calcium	0·015%
of the bladder and	phosphates	0·15%
ureters)	sulphates	0·18%
	and others	—

The specific gravity of this fluid is 1·001–1·040. Its reaction is usually slightly acid in nature with a pH about 6. This reaction really depends upon the diet. The vegetarian diet produces an alkaline pH. Urine passed in the early morning is darker and more concentrated because it contains less water.

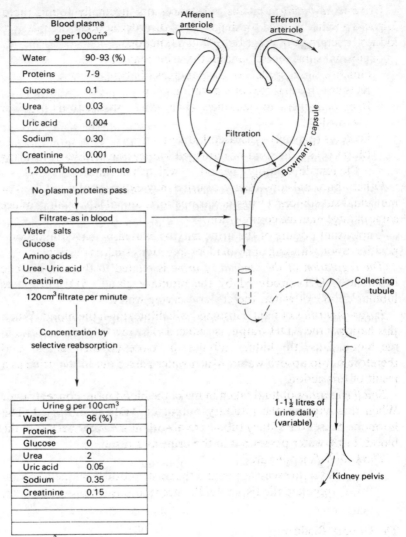

FIG. 55. Urine formation.

Urine in ill-health contains substances not normally found there. Urine is a valuable fluid for diagnosis of disorder in body chemistry or kidney function or of other body organ function.

Abnormal substances found in the urine may be:

Glucose—in disorders of the pancreas (diabetes).

Ketones—in diabetes or starvation.

Bile pigments—in jaundice (bile duct obstruction or liver disorder).

Proteins (plasma)—diseased kidneys.

Blood cells—diseased or damaged kidneys or bladder (or during the female monthly "period"—which is normal).

Other abnormal constituents of urine may indicate coloured food or medicinal substances. Drugs of various sorts and alcohol will produce some change in urine composition.

Unpleasant odours of the urine may be caused by bacterial decomposition upon standing or by bladder infections such as *cystitis.*

The regulation of the amount of urine is related to the anti-diuretic hormone (ADH) produced by the pituitary gland. ADH increases tubular uptake of water, thereby conserving water.

Large quantities of fluid consumed will dilute down the blood. When this happens the ADH output is reduced. The low ADH quantities in the blood cause the kidney tubules to "close up their pores" and therefore not to absorb water. Much water passes out in the urine as a result of this action.

Small quantities of fluid taken in make the blood more concentrated. When this happens the pituitary output of ADH is increased. The hormone causes the kidney tubules to absorb more water back into the blood. Little water passes out in the urine as a result.

The kidney functions are:

(a) To excrete excess water and the waste products of metabolism.

(b) To regulate the tissue fluid concentrations of salts and water.

The Urinary Bladder

This is a hollow muscular bag situated in the pelvic region which is receiving urine continuously in small quantities. It can hold a variable quantity of urine between 120 cm^3–350 cm^3. When the muscle walls become distended with the increasing volume of urine the muscle contracts and expels the urine. Emptying the bladder is called *micturition,* a reflex mechanism which has become largely under voluntary control in the adult.

Urine passes out of the bladder through the *urethra.* This tube is longer in the male, being about 20 cm, since it passes through the prostate gland and down the penis. In the female the urethra is much shorter being only about 4 cm in length. Females are more liable to bladder infections because of this short passage between the bladder and a possible source of infection.

SOME DISEASES AND DISORDERS OF THE URINARY SYSTEM

Bright's Disease. An inflammatory disorder of the kidney not directly related to bacterial infection. Water tends to be retained by the body and protein is lost in the urine.

Calculus. Kidney "stones" commonly made of calcium phosphate or calcium exalate. Often associated with bladder or kidney infection. They can hinder the flow of urine.

Cystitis. An inflammation of the urinary bladder usually caused by bacterial infection. Characterized by frequent and painful passage of urine.

Enuresis. Nocturnal bed-wetting, probably with psychological causes.

Glomerulonephritis. An inflammation of the glomeruli caused by no apparent bacterial infection or poison. Blood and protein links from the glomeruli and appears in the urine.

Nephritis. Inflammation of the kidneys as a response to bacterial infection. The infection may reach the kidneys from the bladder by way of the ureter.

Oedema. An accumulation of water in the tissues. This accompanies many kidney diseases. Water is retained and puffiness of the skin results.

Pyelonephritis. This is an inflammation of the pelvic region of the kidney most commonly caused by bacteria reaching it by way of the ureter.

Uraemia. An excess of urea in blood because of defective kidney function. Urea is not itself poisonous.

CHAPTER 9

The Skin

THE skin is the outer covering of the entire body. It is a vital organ, essential to life and health. It is intimately associated with the structures that lie beneath it. Death will soon follow if large areas of the skin are damaged, such as may result from severe burning. This damage destroys the sweat glands which are important in the temperature regulation of the body.

The skin makes up one-sixth of the total body weight. Somebody has calculated the following:

A 70 kg (11 stone) man who is 170 cm (5 ft 6 in.) tall has a skin area of 16,000 cm². The weight of this skin is 3000 g (6·6 lb) and has a volume of 2400 ml.

The skin is, therefore, a very substantial organ.

THE STRUCTURE OF THE SKIN

The skin is made up of two layers:

(a) *The epidermis* is the outer layer of cells which contain no blood vessels or lymphatics. This protective layer has several layers of cells; it is a stratified epithelium. It varies in thickness depending upon the area of the body; it is very thick and horny in places where much friction takes place such as on the heel of the foot. Raised areas of the epidermis with a serum filled cavity beneath produced by rubbing are called *blisters*.

The layers making up the epidermis are as follows:

Stratum corneum (horny layer, scale like dead cells)—outermost.

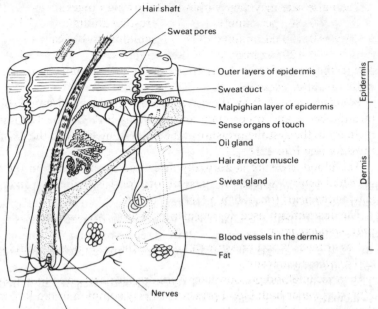

FIG. 56. A diagram of the skin and hair in vertical section.

Stratum lucidum (clear layer, no nuclei, cytoplasm changing to keratin).

Stratum granulosum (cytoplasm granular, nuclei present).

Stratum germinativum (growing layer of cells overlying the dermis)—innermost.

As cells become rubbed away from the surface layer they are replaced by cells growing up from beneath. The growing and multiplying cells of the germinative layer go through a series of changes as they are pushed up to the surface of the skin. The time taken for new cells to pass through all these stages and to become horny surface scales is said to be $2\frac{1}{2}$ to 5 weeks.

Living cells passing up through these stages become *keratinized,* horny dead cells losing their nuclei. This keratinization is an active process not just a dying off. Enzymes located in the granular layer are associated with this change.

Keratin is a protein which makes up most of the hair and dead skin.

The colour given to the skin is caused by a mixture of factors; the nature of the epidermis, the oxygenation or deoxygenation of the blood flowing beneath the epidermis, or the pigment (melanin) granules in pigment cells (melanocytes) of the germinating layer.

Sun tan is caused by the action of ultra-violet radiations falling upon the skin. The pigment-forming materials in the epidermis darken when these radiations play upon the skin. This pigment formation protects deeper layers from the harmful rays of the sun.

Vitamin D is also produced in the skin when UV rays fall upon it.

The surface of the epidermis is a rough protective barrier against water and other materials which might otherwise penetrate the skin. The acid nature of the surface is another defence against bacteria and fungi which might otherwise grow on the skin.

(b) *The dermis* is made up of connective tissue with blood vessels, nerves and lymphatics. It projects upwards into the epidermis as *dermal papillae*. The dermis is elastic in nature because of the elastic fibres running throughout.

Beneath the dermis is a fatty layer of subcutaneous tissue.

Derivatives of the Skin

The following structures are derived from the skin:
  (i) *The Sweat glands.*
 (ii) *The Sebaceous glands* (oil glands).
(iii) *The Hairs.*
 (iv) *The Nails.*

(i) THE SWEAT GLANDS

It has been calculated that the adult body has approximately 2·5 million sweat glands pouring out about 1·5 litres of sweat per day. In very hot air they can pour out as much as 10 litres (2 gallons) of sweat with 30 g of salt in a day. Sweat glands are found in the skin all over the body, large numbers being located in the arm-pits, groins, forehead, palm of the hand and the sole of the foot.

There are two types of sweat glands:

THB—K

The apocrine gland is larger and found in areas near sex organs, arm-pits and groins. This type of gland is derived from the hair follicle. They produce a fluid which can develop a strong unpleasant odour when bacteria on the unwashed skin acts upon the sweat.

The eccrine gland is smaller and is distributed throughout the body. This type of sweat gland is seen in Fig. 56 as a twisted tubular structure lying deep in the dermis. The sweat is produced in the coiled section.

The eccrine glands are under the control of nerves from the sympathetic nervous system whereas the apocrine glands have no nervous control. The apocrine glands are stimulated to action by the hormone adrenaline.

Sweating is described as *insensible* and *sensible*.

Insensible sweating is occurring all the time, unnoticed. Much of this sweat is evaporated before it reaches the skin surface.

Sensible sweating takes place when the body is likely to become overheated because of exercise or a hot atmosphere.

Sweating is not under conscious control.

The composition of sweat is as follows:

Water—98%

Dissolved substances— 2%   sodium chloride
lactic acid
urea (a trace)
glucose
iron (a trace)
ammonia (small amounts)

Sweating is a means of removing water and salts from the body. It is also an important means of regulating the body temperature.

Temperature regulation

Heat produced by the body chemistry must be lost to the surroundings in order to keep man's body at a definite temperature. Man's body temperature is kept at a steady 37°C (98·4°F) by balancing the heat lost with the heat gained. Man is *homoiothermous*, like other mammals which are capable of maintaining a body temperature which is independent of the surroundings.

Heat is gained as follows:
   From oxidation of foods.
   From muscle, liver and other tissue activity (internal metabolism).
   From hot foods or drinks.
If the total heat gain is insufficient some extra heat is generated by involuntary muscular contractions called *shivering*.

Heat is lost as follows:
   From the skin where the blood vessels dilate to allow a greater
      flow of blood.
   Heat is lost by conduction, convection and radiation.
   Sweat is evaporated and the heat needed for this is removed
      from the skin.
   From the lungs, warm air is expired.
   From the urine and faeces (small quantities).
The maintenance of body temperature is controlled from the temperature-regulating centres in the brain, in the *hypothalamus*.

In fever (pyrexia) the temperature-regulating mechanism seems to become upset and the temperature rises to a new figure such as $40°C$ ($104°F$) or even higher.

(ii) SEBACEOUS GLANDS (OIL GLANDS)

Oil is poured out onto the skin surface from glands emptying into the hair follicles. This *sebum* lubricates and protects the hair and skin. Sebum is anti-bacterial and fungicidal in action.

The amount of oil secreted is related to the size of the gland. The size of these glands is related to hormones present in the blood. During adolescence these glands are particularly active.

"Chapped" skin can result from the absence of skin oil allowing water loss from the epidermis.

(iii) THE HAIRS

Hairs are characteristic of the mammal. Man has hair on most parts of the body except in such places as finger-tips, toe-tips, parts of the sex organs, the lips, the soles of the feet and the palms of the hand.

Infant downy *lanugo* hair appears first on the head of the unborn child at about 3–5 months. It is generally unpigmented. The foetal lanugo hair is shed from the face and head between 7–8 months and is replaced by longer stiffer hair on the scalp called *terminal hair*. The scalp has 85 % terminal hairs and 15 % new downy or *vellus* hairs.

Estimates indicate that the average human scalp of 771·6 cm² has about 120,000 hairs, but they are distributed differently. For instance, the crown of the scalp—300 hairs per cm²; the back of the head—200 hairs per cm².

This compares with:
    the beard region—30–40 hairs per cm²,
    the back of the hand—15–20 hairs per cm².

The rate of growth of scalp hairs is about 0·35 mm per day growing to 56 cm or rarely to 94 cm. Hairs on other parts of the body grow slower, at about 0·2 mm per day. Cutting the hair has no effect on these growth rates.

Baldness (Alopecia) is caused by a variety of factors but the commonest, male pattern baldness, is largely hereditary. The causes and cures of baldness are not yet known.

Grey hair does not exist; it is a mixture of white, non-pigmented hair with pigmented hair. People do not go "grey" overnight!

Hair structure

The hair is a dead structure produced from the living cells of the hair bulb surrounding the *dermal papilla*. The hair is sheathed by a tube called the *hair follicle*. Attached to the side of the hair follicle are the *hair erector muscles* (arrectores pilorum) which pull the hair into the vertical position when the skin is cooled. They produce the 'goose flesh' of chilled skin. The oil glands are also attached to the side of the hair follicle.

The hair shaft is made up of the following parts:

The cuticle—overlapping colourless scales. The free edges of these
    scales point towards the tip of the hair.
The cortex—lies on the inside of the cuticle. The cortical cells are
    boat shaped with air spaces between them.
The hair colour pigment is found as small granules within the cortex
    cells.

The medulla—is the central core of the hair. It is not always present along the whole length of the hair shaft. The air spaces within the medulla are partly responsible for the sheen of the hair in bright light. Some colour is also present in the medulla.

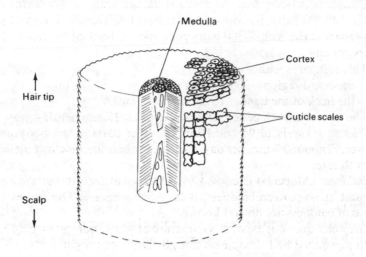

FIG. 57. A section of the hair shaft.

The way in which a hair begins life in the skin is shown in Fig. 58.

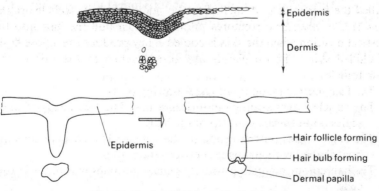

FIG. 58. The hair follicle and bulb forming.

(iv) THE NAILS

The last joints of the fingers and toes are protected by translucent horny plates called the *nail plates*. The clear layer of the epidermis forms the nail, the cornified layer of the epidermis forms the thin cuticle.

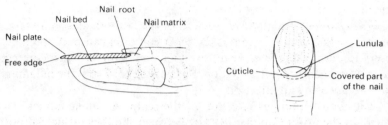

FIG. 59. The nail.

The nail plate rests upon an extremely sensitive *nail bed*. The plate is intimately linked with the bed below from which is receives its nutrients. When the nail plate becomes separate from the nail bed the whole area becomes discoloured and distorted.

The fixed end of the nail shows a half moon or *lunula*. This structure is where the nail bed becomes the *nail matrix*. It is the nail matrix which forms the hard and horny cells of the nail plate.

When the nail grows it moves forward with the nail bed. There is no limit to the growth of a nail, it grows about 3 mm (1 in.) per month. The greater growth is on the nail of the third digit.

SOME DISEASES AND DISORDERS OF THE SKIN

Abscess. This is a collection of pus in a subcutaneous cavity or elsewhere. Usually the result of staphyloccal infection.

Acne vulgaris. A skin disorder related to the abnormal activity of the skin oil glands. Associated with adolescence.

Boil. This is a small abscess around the root of a hair or a sweat gland. It is not generally a good idea to squeeze boils.

Comedo (a blackhead). A plug of greasy material in the opening of an oil gland.

Carbuncle. A multiple boil. Deeper and larger than the single boil.

Eczema. A dermatitis of an inflammatory nature which forms blisters and skin scales. The causes are numerous, for example a skin response to irritants or an hereditary tendency to a faulty cellular metabolism.

Gangrene. The widespread death of body tissues. The usual cause is the loss of blood supply because of injury or disease of the artery in the area.

Inflammation. A defensive reaction of the body tissues to irritation or injury. Bacterial infection is the most common cause of inflammation. Redness, soreness, heat, swelling and pain are the common accompaniments of inflammation. Pus is often formed if the inflammation is caused by an infection.

Ringworm. A fungus infection of the scalp. A similar attack may occur in the groin ('dhobie itch) or between the toes (athlete's foot). The nails of the fingers may also be attacked by this fungus.

CHAPTER 10

The Brain and Nerves

THE NERVOUS SYSTEM

It is necessary to communicate between one part of the body and another. This communication can be rapid or slow, over many months or even years. The nervous system is the rapid communication system, the endocrine glands with hormones are the slower communication systems. These are studied later. The nervous system will be studied under the following headings:

    (a) *The Central Nervous System* (CNS)
       (i) *The brain and twelve pairs of cranial nerves*
      (ii) *The spinal cord and thirty-one pairs of spinal nerves*
    (b) *The Autonomic Nervous System* (ANS)
       (i) *The sympathetic nervous system*
      (ii) *The parasympathetic nervous system*
    (c) *The Nerve Impulse.*

(a) THE CENTRAL NERVOUS SYSTEM

(i) The Brain and Cranial Nerves

The brain and cranial nerves are enclosed within the skull. The cranial nerves leave the brain and pass through the skull in order to reach various organs.

THE BRAIN

The brain has the following main parts:
The forebrain. Large *cerebral hemispheres (the cerebrum)* which

make up the greatest part of the brain. The outer region of the hemispheres is a layer of *grey matter* making up the *cerebral cortex* (1·5–4·5 mm in thickness).

The mid-brain. The forebrain arises from this stem. It contains a criss-cross formation of white and grey matter. It is a communication centre for one part of the brain with another.

The hind-brain. The mid-brain meets the hind brain at the *pons Varoli* which is a thickened stem portion facing forwards. The *cerebellum* is facing backwards and is opposite the pons. The *medulla oblongata* is the lower part of the brain-stem which continues as the spinal cord after it leaves the skull.

The cells making up the parts of the brain just mentioned are called *neurons*. These nerve cells have been studied at an earlier stage. The so-called *grey matter* is the description of the nerve cell bodies, the *white matter* is made up of nerve fibres covered in white fatty sheaths. The outside of the brain has many folds (convolutions) of grey matter thereby increasing the number of nerve cells present. The inside of the brain is mostly nerve fibres, white matter.

Some of the brain parts just described have the following structures and functions:

The cerebrum (cerebral hemispheres)

This is the largest part of the brain made up of two hemispheres. The outer part of the cerebral hemispheres is made up of a much folded or convoluted cerebral cortex. The grey matter making up the cerebral cortex is largely nerve cells. Nerve fibres run to and from the nerve cells in the cortex and they make up the inner white matter of the brain hemispheres. (Note the reverse situation in the spinal cord: white matter overlies the grey matter.)

The cerebral hemispheres may be divided into four lobes:

Frontal lobe.
Parietal lobe.
Temporal lobe.
Occipital lobe.

The frontal and parietal lobes are separated by a deep cleft called a fissure. It is called the *fissure of Rolando*.

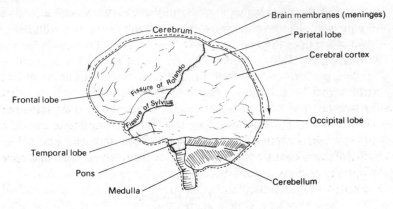

FIG. 60. Parts of the brain.

The temporal lobe lays beneath another cleft called the *fissure of Sylvius.*

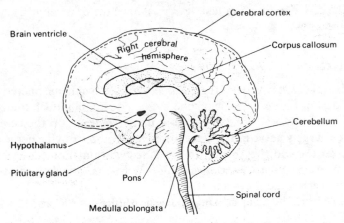

FIG. 61. Vertical section through the brain.

The two hemispheres are separated by a deep cleft, the *longitudinal fissure.* At the base of this cleft the left and right hemispheres are linked together by way of fibres in the *corpus callosum.*

Within each cerebral hemisphere is a series of hollow cavities called brain *ventricles*. These ventricles contain a fluid, *cerebrospinal fluid*.

Underneath the cerebrum near the mid-brain is the *hypothalamus,* from it is suspended the *pituitary gland.*

The functions of the cerebrum are localized in certain areas of the cerebral cortex. This "mapping out" of the areas on the cerebral cortex has been done by electrical stimulation of various parts and noting the muscular and sensory responses. The muscles that respond are on the side of the body opposite to the hemisphere stimulated. The fibres travelling to and from the cortex cross over from left to right and vice versa.

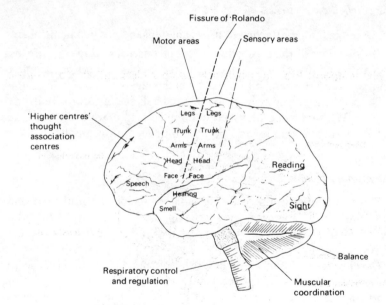

FIG. 62. Locality of brain functions.

Motor areas are those parts of the cerebral cortex just in front of the Rolando fissure. Impulses from these areas activate (motorize) muscles in the structures mentioned in Fig. 62.

Sensory areas are those parts of the cerebral cortex just behind the Rolando fissure. Electrical stimulation of different parts of this area cause the conscious person to describe certain sensations. The localities of the sensory areas are shown in Fig. 62.

The *frontal lobes* are largely unmapped and are probably concerned with the "higher" mental processes, intelligence, memory, conscious thought, etc.

The *hypothalamus* has important regulatory functions for touch, temperature, blood pressure and emotional responsiveness.

The *pituitary gland* is often called the "master gland". It secretes a series of very important hormones studied in a later chapter.

The cerebellum

The cerebellum is attached to the brain stem and has two hemispheres.

The *functions* of the cerebellum are "unconscious" and our information about its importance depends upon studies of injured people or animals where the cerebellum has been damaged.

The cerebellum is necessary for normal balance and coordination of muscles. Without a complete cerebellum the person walks in a drunken fashion and shows shaky and poorly controlled movements.

The medulla oblongata

This is a continuation of the spinal cord inside the skull. It is here that some nerve fibres cross over from one side of the body to the other. Here the white matter is on the outside, the grey matter on the inside, as in the spinal cord.

The *functions* of the medulla are also "unconscious" because the autonomic reflex control centres are located here. The respiration, heart rate and blood pressure regulation originates in this area. Swallowing, coughing, sneezing and vomiting reflexes are integrated in the medulla.

The brain membranes (meninges)

The *meninges*—enclosing the brain and spinal cord are:
    The dura mater—next to the bone usually.

The arachnoid mater—middle membrane.

The pia mater—next to the nervous tissue.

Between the arachnoid and pia there is a fluid, the *cerebro-spinal fluid*, in the *sub-arachnoid space.*

The dura mater is a tough membrane lining the inside of the cranium and spinal canal. It dips in and out of the large clefts in the brain helping to support the bulk of the cerebral hemispheres and the cerebellum.

The arachnoid mater lies beneath the dura mater. It is a flimsy membrane attached to the undersurface of the dura mater.

The pia mater is a delicate membrane adhering to the brain and spinal cord surfaces. It carries blood vessels.

The cerebro-spinal fluid is found between the arachnoid and pia acting as a support for the brain and spinal cord.

Removal of this fluid may be necessary for diagnostic purposes by a *lumbar puncture,* which means drawing off some fluid by a needle inserted into the sub-arachnoid space in the lumbar region of the spine.

Brain waves

The nerve cells of the brain generate small electric currents which can be picked up by electrodes, magnified and recorded. The electrical events can be seen as waves on a cathode-ray tube or as pen traces on moving paper.

These "brain waves" were first demonstrated in 1875 by an Englishman, Caton, using very primitive equipment and working with dogs. Human brain studies by an Austrian, Berger, between 1929 and 1938 revealed some interesting factors about these electrical brain rhythms, later called *Berger rhythms.* The recording of these rhythms are called *electroencephalograms* (EEG). The origins of this electrical activity in the brain is obscure. When the eyes are closed a rhythm called an *alpha rhythm* can be recorded (it is 8–13 cycles per second). When the eyes are opened or the person does some mental work such as some arithmetic (with the eyes closed), the alpha rhythm disappears.

There are other waves recorded from the brain such as those recorded when a person is asleep, called the *delta rhythm* ($\frac{1}{2}$–3 cycles per second).

Studies of these rhythms and their changes in frequency can be of great diagnostic value in identifying types of epilepsy or some other brain disorder.

THE CRANIAL NERVES

There are twelve pairs of nerves leaving the brain.
Some of these nerves carry motor impulses only *(motor nerves)*.
Other nerves carry sensory impulses only *(sensory nerves)*.
Some nerves carry both types of impulse *(mixed nerves)*.
A motor nerve carries impulses *from the brain to* a muscle or organ.
A sensory nerve carries impulses *from the organ or tissue* to the brain.
Mixed nerves carry both of these impulses.
The cranial nerves and their functions are summarized in Table 8.

TABLE 8

S = sensory nerve   M = motor nerve   MS = mixed nerve

No.	Nerve	Type of nerve	Part supplied	Function
I	Olfactory	S	Nasal epithelium	Sense of smell
II	Optic	S	Retina of the eye	Vision
III	Oculomotor	M	Four eyeball muscles	Eyeball movement
IV	Trochlear	M	One eyeball muscle	
V	Trigeminal	S	Face, scalp, nose, teeth	Sensations from these areas
		M	Jaw muscles	Chewing action
VI	Abducens	M	One eyeball muscle	Eyeball movement
VII	Facial	S	Tongue, soft palate	Taste
		M	Face muscles	Facial movements
VIII	Auditory	S	Ear	Hearing Balance
IX	Glossopharyngeal	S	Tongue, pharynx	Taste
		M	Throat muscles, parotid gland	Throat movements
X	Vagus	S	Heart, lungs, digestive tract, etc.	Sensations from these organs
		M	Heart, lungs, digestive tract, etc.	Movements of these organs
XI	Accessory	M	Muscles of neck and shoulders	Movements of these areas
XII	Hypoglossal	M	Muscles of the tongue	Movements of the tongue

Twelve pairs of nerves arise directly from the under surface of the Brain to supply Head and Neck and most of the viscera.

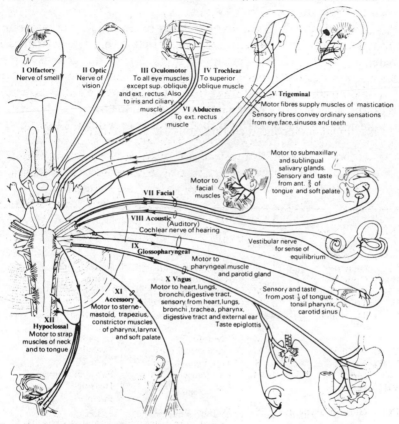

I Olfactory
Nerve of smell

II Optic
Nerve of vision

III Oculomotor
To all eye muscles
except sup. oblique
and ext. rectus. Also
to iris and ciliary
muscle

IV Trochlear
To superior
oblique muscle

VI Abducens
To ext. rectus
muscle

V Trigeminal
Motor fibres supply muscles of mastication
Sensory fibres convey ordinary sensations
from eye, face, sinuses and teeth

Motor to submaxillary
and sublingual
salivary glands.
Sensory and taste
from ant. $\frac{2}{3}$ of
tongue and soft palate

VII Facial
Motor to
facial
muscles

VIII Acoustic (Auditory)
Cochlear nerve of hearing

Vestibular nerve
for sense of
equilibrium

IX Glossopharyngeal
Motor to
pharyngeal muscle
and parotid gland

X Vagus
Motor to heart, lungs,
bronchi, digestive tract,
sensory from heart, lungs,
bronchi, trachea, pharynx,
digestive tract and external ear
Taste epiglottis

Sensory and taste
from post $\frac{1}{3}$ of tongue,
tonsil pharynx,
carotid sinus

XI Accessory
Motor to sterno-
mastoid, trapezius,
constrictor muscles
of pharynx, larynx
and soft palate

XII Hypoglossal
Motor to strap
muscles of neck
and to tongue

FIG. 63. Cranial nerves. (Reproduced by permission from Nurses' Illustrated Physiology by Mc. Naught and Callender, E & S, Livingstone Ltd., Edinburgh.)

(ii) The Spinal Cord and Spinal Nerves

The spinal cord is continuous with the medulla oblongata and continues below the skull down the spinal canal within the backbone. The spinal cord is an extension of the brain and is covered by the same membranes or meninges.

The spinal cord in man ends at a point just below the 12th ribs, at the first lumbar vertebra. Below this point the nerves travel down the spinal canal, looking rather like a horse's tail and are so named the *cauda equina*.

The spinal cord is a column of nervous tissue connecting the brain with the rest of the body. It is the spinal cord that takes a part in reflex action without reference to the brain. Such reflex actions as the knee jerk require the spinal cord only. Notice that the white matter (containing nerve fibres travelling to and from the brain and elsewhere) is on the outside of the cord. Grey matter with the nerve cell bodies is on the inside of the cord. This is the other way round in the brain.

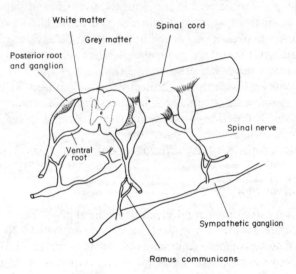

White matter

Grey matter

Spinal cord

Posterior root and ganglion

Spinal nerve

Ventral root

Sympathetic ganglion

Ramus communicans

FIG. 64. Spinal cord and spinal nerves in section.

The H-shaped grey matter contains nerve cells which link up various nerve fibres as follows:

In the posterior horns of this grey matter cells *synapse* (do not join but come close together) with incoming nerve fibres whose cell bodies lie outside the spinal cord in the *posterior root ganglia*. These carry sensory impulses into the spinal cord. They are called *afferent or sensory neurones*.

In the anterior horns of the grey matter there are the nerve cell bodies of the outgoing fibres carrying motor impulses to mobilize voluntary muscles. They are called *efferent or motor neurones.*

A *connector or intermediary neurone* links up the posterior and anterior horns of the grey matter.

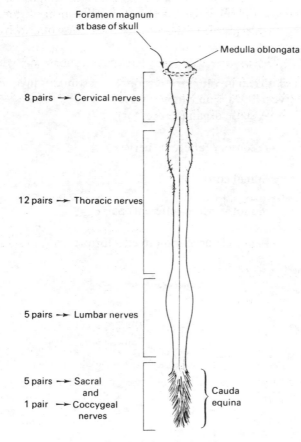

Fig. 65. The spinal cord.

The spinal nerves are thirty-one pairs of nerves emerging from the spinal cord and corresponding to the vertebrae of the backbone. The nerves leave the backbone through spaces between a pair of vertebrae.

These nerves carry motor and sensory neurones; they are mixed **nerves** grouped as follows:

Cervical nerves	8 pairs
Thoracic nerves	12 pairs
Lumbar nerves	5 pairs
Sacral nerves	5 pairs
Coccygeal nerves	1 pair

THE SPINAL REFLEX

A spinal reflex is an involuntary response to a stimulus involving the following nerve pathways and structures:

A sense organ (a receptor)
↓
A sensory (afferent) nerve
↓
Spinal cord
↓
A motor nerve (efferent) nerve
↓
A muscle or gland (an effector organ)

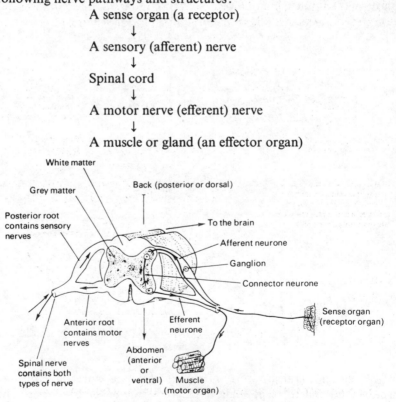

FIG. 66. Reflex arc.

An example of a *reflex action* is the *knee jerk* produced by tapping the patella tendon with a rubber hammer. The quadriceps muscle of the thigh suddenly contracts because of this stimulus and the leg kicks forward. This demonstration is carried out when the leg is crossed over the other, in the sitting position. This is unlearned, involuntary behaviour.

More natural reflex actions are blinking the eye in response to a stimulus such as a quick movement or a piece of grit flying towards the eye, coughing and sneezing when the mucosa is irritated. Vomiting, swallowing, sweating in response to heat and deep breathing because of the build up of carbon dioxide in the blood are all reflex actions to a particular stimulus, but not all are restricted to the spinal cord because some involve the medulla of the brain (respiratory centres).

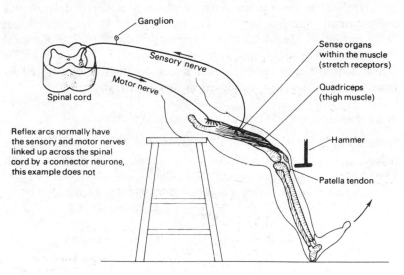

FIG. 67. The knee-jerk reflex.

The *reflex arc* is the physical basis of a reflex action and consists of the organs and nerves as mentioned previously and as shown in Fig. 67.

Some characteristics of reflex action:

(a) They are involuntary and control many of our important functions such as respiration, digestion, etc. Much of our voluntary activity contains an element of reflex action such as climbing upstairs, riding a bicycle, emptying the bladder.
(b) Reflexes are usually quick responses to a stimulus and prevent the body from being damaged.
(c) They usually involve the spinal cord only but the "higher centres" of the brain can influence reflex action as can the autonomic nervous system. Learning can influence the emptying of the bladder as can emotion.
(d) Reflex actions may be *inborn* or *conditioned*. Conditioned reflexes are modifications of inborn reflexes, or new automatic actions learnt from experience. Most of our bodily functions are controlled by the inborn reflexes which are inherited from our parents. Conditioned reflexes are acquired by each individual during his life. Much of our learning is the acquiring of conditioned reflexes.

The acquisition of conditioned reflexes was first demonstrated by Ivan Pavlov in an article published in Russia in 1906.

THE EXPERIMENTS OF PAVLOV ON CLASSICAL CONDITIONING

Russian physiologist, Ivan Pavlov (1849–1936), studied reflex behaviour in dogs. He noticed that a dog will salivate when shown meat. The *sight* not the taste of meat is causing the flow of saliva. The dog presumably *learnt to associate* the sight of meat with the eating of meat.

$$\text{Stimulus } S^1 \longrightarrow \text{Response } R^1$$

| (eating meat) | (salivation) |
| (unconditioned stimulus) | (unconditioned response) |

After some experience with meat a dog learns to associate the sight of meat with eating meat.

$$\text{Stimulus } S^2 \longrightarrow \text{Response } R^1$$

| (sight of meat) | (salivation) |
| (conditioned stimulus) | (unconditioned response) |

Pavlov's experiments demonstrated that saliva flow can be caused by other conditioned stimuli, a bell, a flashing light, or a ticking metronome.

The dog is presented with a flashing light just before having meat powder blown into its mouth. After many trials the dog salivates when only a light flashes.

$$\text{Stimulus } S^1 \longrightarrow \text{Response } R^1$$
$$\text{(unconditioned)} \qquad \text{(unconditioned)}$$
$$\text{Meat} \qquad \text{Salivation}$$

$$\text{Stimulus } S^2 \xrightarrow{\text{Many Trials}} \text{Stimulus } S^1 \longrightarrow \text{Response } R^1$$
$$\text{(conditioned stimulus)} \qquad \text{(unconditioned)} \qquad \text{(unconditioned)}$$
$$\text{Flashing light} \qquad \text{Meat} \qquad \text{Salivation}$$

After repeated pairings of light with meat the dog will salivate to the flashing light; this is a conditioned response.

$$\text{Stimulus } S^2 \longrightarrow \text{Response } R^2$$
$$\text{(conditioned} \qquad \text{(conditioned}$$
$$\text{stimulus)} \qquad \text{response)}$$
$$\text{same as } R^1$$
$$\text{Flashing light} \qquad \text{Salivation}$$

This R^2 will cease if S^1 (meat) is not given frequently enough; the conditioned response must be rewarded or *reinforced*. With the absence of a rewarding factor, such as meat, the conditioned response R^2 will become *extinguished*.

This classical conditioning is a very simple form of learning and has been studied with many animal groups including human beings.

A researcher in the study of human behaviour, John B. Watson, demonstrated that an infant's emotional response may be linked to a conditioned stimulus. He showed that a white rat can cause fear in a child if always paired with frightening noises. The rat alone caused no fear.

THE EXPERIMENTS OF SKINNER ON OPERANT CONDITIONING

Professor B.F. Skinner, an American working on behaviour, looked at learning in another way from Pavlov. He rewarded behaviour which was not *normally* brought about by the presence of a particular stimulus. He rewarded a rat with food if it pressed a lever. Rats do not normally press levers when they see them, so in this way operant conditioning differs from classical conditioning. The lever pressing was accidental at first but became part of the rats learned behaviour when that behaviour was regularly rewarded by food.

These experiments are carried out in ingenious electronic "Skinner boxes" or special cages with food-delivery apparatus.

An application of Pavlovian conditioning is used in an attempt to prevent continued drinking in the long-term alcoholic. If a drug which makes one sick and very ill is given along with alcohol then the two become associated, and with luck alcohol drinking will be regarded with distaste.

(b) THE AUTONOMIC NERVOUS SYSTEM

This section of the nervous system controls many functions that are not normally in our consciousness. It is concerned with maintaining an *internal steady state* (homeostasis). It influences the action of the heart, digestive glands, gut movements, muscular movements of blood-vessel walls, etc.

This involuntary section of the nervous system has two distinct sets of nerves as follows:

(i) The Sympathetic Nervous System

Nerves of this system have their origins in the grey matter of the spinal cord. They leave by the anterior (ventral) root with the motor nerves of the spinal nerves. They then synapse with nerve cell bodies in ganglia along the length of the spinal cord, the *sympathetic chain* (see Fig. 68).

The nerve before the ganglion is called the *ramus communicans;* it is a myelinated nerve, white in appearance. The nerves leaving ganglia are grey, non-myelinated nerves.

The nerve endings of most sympathetic fibres liberate a chemical transmitter called *noradrenaline*. These fibres are described as *adrenergic*. This chemical stimulates the target organ into action. The released chemical is quickly taken back into the nerve endings afterwards.

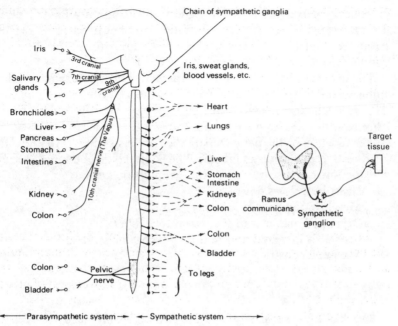

FIG. 68. The autonomic nervous system.

The function of the sympathetic nervous system is to initiate the emergency actions of "flight and fight". The bodily changes taking place are as follows:

Heart beat speeds up.

Muscle blood vessels dilate allowing more blood to flow.

Respiratory tubes increase in diameter.

Peristalsis slows down; gut enzyme secretion is reduced.

The pupil dilates.

These changes allow the body to act at a more rapid pace in an emergency. The accompanying sensations are described as an emotion of fear or excitement.

(ii) The Parasympathetic Nervous System

The nerves of this system are similar to the previous in having a myelinated nerve before the ganglion and a non-myelinated nerve after the ganglion. In this case the ganglia are not lined up as a chain, as in the sympathetic system. The ganglia are situated *in the organ* supplied.

The nerve endings of the parasympathetic release a chemical transmitter called *acetylcholine*. These fibres are described as *cholinergic*. The acetylcholine is inactivated by an enzyme called *acetylcholineesterase* with splits it up after it has transmitted the nerve impulse.

The function of the parasympathetic nervous system is to allow the body rest, recuperation and repair. The bodily changes taking place are the opposite of those caused by the sympathetic system.

Heart beat slows down.

Muscle blood vessels constrict.

Decreased diameter of the respiratory tubes.

Peristalsis increases and so does gut enzyme secretion.

The pupil constricts.

(c) THE NERVE IMPULSE

The living nerve at rest has an electrical difference between its outside and inside. This is because sodium ions are "pumped" out of the axon cytoplasm and this makes the inside of the nerve negative when compared with the outside. Potassium ions are in higher concentration within the axon fibre.

When the membrane is stimulated by some means or other the membrane permeability to potassium ions changes and they move out of the nerve. Sodium ions move inwards. These movements of ions are the nerve impulse and it spreads along the length of the nerve rather as the spark moves down a lighted fuse. The resting condition of the nerve is rapidly restored when the mysterious "sodium-pump" begins to eject sodium ions across the nerve membrane once more.

When a nerve impulse reaches a synapse or muscle it liberates a *nerve impulse-transmitting chemical* called acetylcholine. This chemical moves across the small separating space and stimulates the membrane of the next nerve or the muscle into activity.

Nerve-muscle preparations may be studied by using a rotating recording drum called a *kymograph*. The muscle and nerve are taken

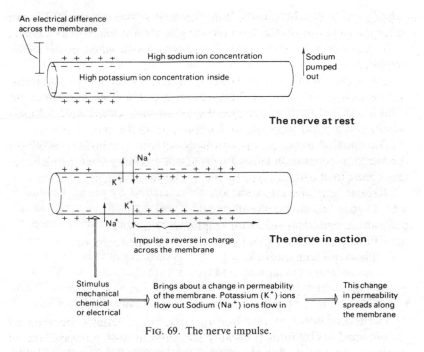

An electrical difference across the membrane

+ + + + +
- - - - -
High sodium ion concentration
High potassium ion concentration inside

Sodium pumped out

The nerve at rest

Na$^+$
K$^+$
K$^+$
Na$^+$

Impulse a reverse in charge across the membrane

The nerve in action

Stimulus mechanical chemical or electrical ⟶ Brings about a change in permeability of the membrane. Potassium (K$^+$) ions flow out Sodium (Na$^+$) ions flow in ⟶ This change in permeability spreads along the membrane

FIG. 69. The nerve impulse.

from a recently killed frog. The nerve is stimulated by electricity from an accumulator and the muscle actions are recorded by a moving ink pen tracing on a rotating paper.

SOME DISEASES AND DISORDERS OF THE NERVOUS SYSTEM

Bell's Palsy. A paralysis of the facial nerve (fifth cranial nerve). The face muscles no longer being able to move, produce the one-sided lack of expression typical of this disorder.

Disseminated sclerosis (multiple sclerosis). A chronic disease of the central nervous system. There is some discussion suggesting that long term virus activity may be involved in the disease. Scattered areas of the brain and spinal cord degenerate and nerve fibres lose their insulating sheath of fat (myelin sheath). The nerves lose their ability to conduct impulses.

Epilepsy. A general name for nervous disorders characterized by fits. A tendency toward epileptic fits may be hereditary.

Herpes. There are two types of herpes virus infection, Herpes simplex producing cold sores and, Herpes zoster giving rise to shingles and chicken-pox.

Lumbago-sciatica. A continuing pain in the small of the back. It is often associated with sciatica pains running down the leg. It may be brought on by "slipped disc" or by faulty posture.

Meningitis. A virus or bacteria infection of the brain membranes, the meninges. An examination of the cerebro-spinal fluid is one way of diagnosing meningitis.

Migraine. An allergic condition characterized by severe headache with no apparent cause. Vomiting and visual disturbance may accompany these episodes which may vary in frequency from person to person.

Parkinson's Disease (Shaking palsy). A disorder of voluntary movement. Involuntary movements and tremors are characteristic. The area of the brain where this control of voluntary action is "regulated" can be influenced to restore normal voluntary activity (to some extent) by a drug L. Dopa.

Poliomyelitis (Infantile paralysis). A virus infection of the grey matter in the front part of the spinal cord. If the nerve cells that supply muscles are damaged by this infection then the corresponding muscles will no longer function. Paralysis results. Vaccines are available to counteract this virus. The Salk vaccine is injected, the Sabin vaccine is given by mouth.

CHAPTER 11

Sense Organs

Our knowledge of the external world comes to us through our sense organs. The organs which receive the various stimuli are grouped as follows:

Exteroceptors—perceive stimuli from the external world.
The eye—perceives visible rays.
The ear—perceives sound vibrations.
The skin—perceives pressure waves and heat rays.
The nose and tongue—perceive chemical changes.
Proprioceptors—perceive stimuli caused by changes in body position.
The ear—perceives movements of the head (semicircular canals).
The muscles and tendons—perceive tension and stretch.
Interoceptors—perceive stimuli from the internal world, such as the
alimentary canal.

THE EYE AND VISION

The Structure

The eye is a sphere about 2·5 cm in diameter with a transparent bulge at the front and an optic stalk at the back. The whole organ is situated within an orbital cavity, attached there by six small muscles. These muscles move the eyeball.

The eyeball is a tough structure with three coats:

An outer sclerotic (sclera) coat. This is a tough fibrous tissue making up the white of the eye. It is protective in function. The front part of the sclera is transparent being called the *cornea*. A thin *conjunctiva* membrane runs over the cornea and lines the inside of the eyelids.

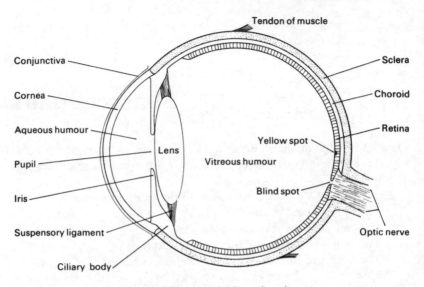

FIG. 70. The eye—a horizontal section.

A middle choroid coat. This is a thin pigmented layer supplied with many blood-vessels. The front region of this layer forms a pigmented muscular ring, the *iris*. This muscular circle can dilate (open) or constrict (close) as a reflex action to the presence or absence of light. The light passes through the *pupil* or hole in the iris. Darkness causes the pupils to dilate thereby allowing more light to enter the eye.

The *ciliary muscles* hold the *lens* in position by means of the *suspensory ligaments* and are responsible for the change in lens shape when the eye focuses. The pigmentation of the iris gives the colour to the eye.

An inner retina coat. This lines the inside of the back of the eye. It is a layer of nervous tissue containing the light-sensitive receptors, the *rods and cones*. A point on the retina, where the nerve fibres pass down the optic nerve on their way to the brain, is known as the *blind spot*. It is insensitive to light as it has no rods or cones. Blood-vessels leave and

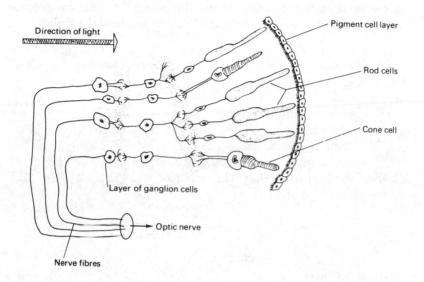

Direction of light

Pigment cell layer

Rod cells

Cone cell

Layer of ganglion cells

Optic nerve

Nerve fibres

FIG. 71. The layers of the retina.

enter here also. Another point on the retina is known as the *yellow spot (fovea centralis* or *macula lutea)* this is the area of greatest, acute vision. It has only cone receptors here for colour reception.

The space between the cornea and the lens is filled with a watery, *aqueous humour*. This fluid is formed by the ciliary process of the ciliary body. The fluid is at a pressure of 25 mm Hg, any increase in this pressure can cause an internal fluid tension known as *glaucoma*. The rest of the eye space behind the lens is filled with a jelly-like *vitreous humour*.

·Moving the Eye

The eyeballs are moved by three pairs of small muscles joining the sclerotic coat to the inside of the eye sockets. These (extrinsic) muscles move both eyes together so that both are looking at the object and the images fall onto similar parts of both retinae.

Protecting the Eye

The exposed part of the eye is protected by the following structures:

The *eyelids* and *eyelashes* protect the eyes against foreign bodies which may otherwise enter the eye. The insides of the eyelids are lined by a thin membrane, the conjunctiva. They distribute lubricating fluid over the corneal conjunctiva as the eyelids blink.

The lacrimal apparatus produces tears in the *lacrimal glands* (tear glands) and so lubricates the exposed eye surface. Tears contain an enzyme which attacks bacteria. The tears are drained away from the eye by way of *lacrimal ducts* which may be seen as minute openings on the inner edges of the eyelids. The tears pass into the back of the nose by way of the nasal duct.

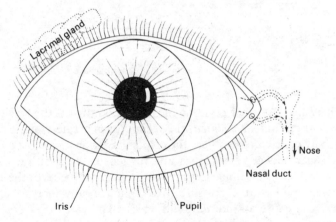

FIG. 72. The lacrimal apparatus.

The unexposed part of the eyeball is protected within the bony eye socket by a lining of fatty connective tissue which prevents damage of the eyeball against the back of the socket.

The Functioning of the Eye Lens

Light passes through the transparent cornea and enters the aqueous humour. This curved surface focuses light onto the lens suspended between the ciliary muscles. This lens can alter its shape and thus vary its focal length.

The action of the lens brings light rays to a sharp focus on the retina. The image is upside down. We learn to "see" things the "right way up" from birth. The ability of the lens to accommodate for near or distant vision results from its changing shape, bringing the image into clear focus on the retina, despite the different distances.

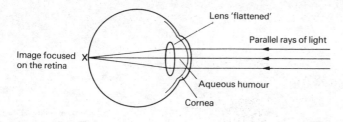

Distant Vision

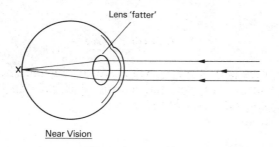

Near Vision

FIG. 73. Near and distant vision.

Distance vision has the lens in the thin, flat condition. The eye at rest is focused for distance.

The lens is flat because of the tension produced by the ciliary muscles pulling on the suspensory ligaments.

Near vision has the lens in the fat, more convex shape. The ciliary muscles contract and relax the tension of the suspensory ligaments on the lens. The elastic capsule of the lens swells because of the absence of tension and the lens becomes fatter.

Disorders of vision can result from having an eyeball the wrong size for the power of the cornea and lens.

Short-sighted people have an eyeball that is too long; *long-sighted* people have an eyeball that is too short.

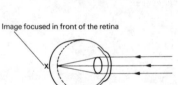

Eye disorder (Myopia)   Correction

Image focused in front of the retina   Image on the retina

Concave lens bends the rays outwards a little

Eye ball is too long

Short sightedness (Myopia)

Eye Disorder (Hypermetropia)

Image focused behind the retina   Image on the retina

Convex lens bends the rays inwards a little

Eye ball is too short

Long sightedness (Hypermetropia)

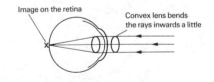

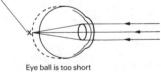

FIG. 74. Short and long sightedness.

The Functioning of the Retina

The retina nerve endings are of two kinds; *rods* and *cones.*

Rods contain a pigment *rhodopsin* (visual purple) which in *dim light* gradually breaks down to *retinene + opsin.* It is the retinine which stimulates nerve impulses to pass to the brain. In bright light this visual pigment is rapidly "bleached" and so when we move from a sunlit street into a dark cinema one is temporarily blinded because the dim vision rods are out of action for about half an hour. In the dark the rhodopsin is resynthesized, in doing so vitamin A is one necessary component. Night-blindness is an early symptom of vitamin A deficiency.

Cones are of three types. The photosensitive pigments which they contain are not known at this time.

Each type of cone has a pigment which is sensitive to a particular colour range.

"Red cones" absorb yellow-orange light.

"Green cones" absorb green light.

"Blue cones" absorb blue light.

The cones function in bright light and so we lack colour vision in dim lighting.

The cones are more numerous around the yellow spot but become less numerous towards the edge of the retina. Rods are more numerous towards the edge of the retina.

Colour blindness is the inability to distinguish certain colours. The different types of colour blindness may be explained as an absence of one or more of the colour cones. With no red cones one is unable to distinguish red from green.

THE EAR, HEARING AND BALANCE

The Structure

The ear is an organ of hearing and balance; it is within the skull. The outer visible part is the *auricle* or *pinna.* There are three main parts to the ear.

The external ear includes the pinna which collects and directs sound waves. The sounds pass down the *external auditory meatus* a short

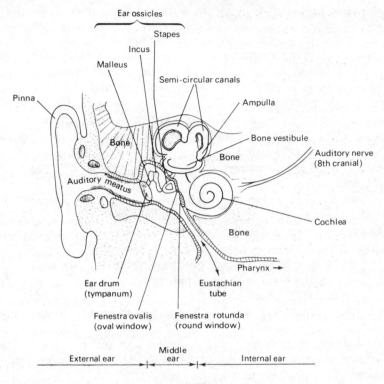

Fig. 75. The ear.

3-cm tube leading to the ear drum. This meatus is lined by wax (cerumen) glands which are modified sweat glands.

The middle ear is separated from the external ear by the fibrous *tympanum* (ear drum). The tiny cavity of the middle ear has crossing it three minute bones, the *ear ossicles*.

These ossicles transmit the vibrations of the ear drum across the middle-ear cavity to the oval window of the inner ear. The middle-ear is air-filled and communicates with the outside air by way of the *eustachian tube*. This tube opens and lets in air when we swallow or sneeze so equalizing the pressures on both sides of the ear drum. It is particularly important to do this when going down in a lift shaft for instance.

The inner ear contains the organs of hearing and balance. The vibrations set up in the three small bones or ossicles (*malleus, incus* and *stapes*) are here transmitted into the fluid which fills the inner ear. This fluid fills a series of complex tunnels or *labyrinths* within bone. Vibrations within this snail-like *cochlea* cause sensitive hairs of the hearing organ, *the organ of Corti,* to move. Nerve impulses leave the cochlea for the brain. Changes in the position of the head are registered in another part of the inner ear labyrinth, in the *vestibular apparatus.* This apparatus consists of the *utricle, the saccule* and the *semicircular canals.*

Hearing

The air is set into motion by banging, talking or whatever. These motions are called sound waves and they have high or low frequencies. The ear is able to distinguish between different frequencies, that is we are able to hear different frequencies of sound (or different *pitch*).

The human ear is able to hear sounds between 20,000 and 20 cycles per second or hertz.

Sound varies not only in frequency but also in loudness, that is in intensity. If the sound is very intense we not only hear it, but also *feel* the vibrations.

The organ of hearing is the organ of Corti within the coiled cochlea. The vibrations in the air are transmitted to the fluid within the cochlea and these vibrations are registered by the organ of Corti. The auditory nerve (8th cranial nerve) carries the nerve impulses to the brain.

The way in which we hear sounds may be summarized as below.

(a) Air vibrations are collected and directed by the pinna into the meatus.

(b) The tympanum sets up vibrations which move the ear ossicles in sympathy. These tiny ear bones transmit the vibrations across the middle ear to the oval window.

(c) The oval window vibrates and moves the fluid within the cochlea.

(d) The movements travelling through the fluid are caused by different frequency sounds. Low-frequency fluid waves travel further down the cochlea than do high-frequency waves. Sounds of similar frequency die out at a similar point along the length of

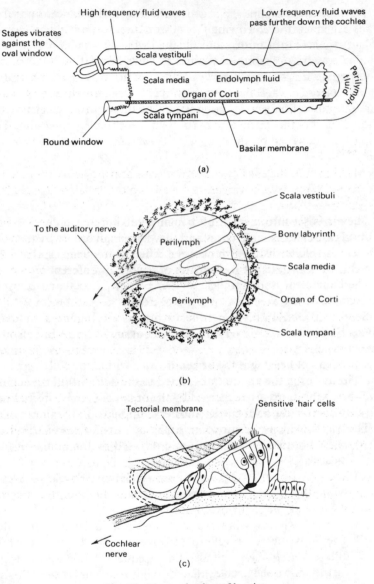

High frequency fluid waves

Low frequency fluid waves pass further down the cochlea

Stapes vibrates against the oval window

Scala vestibuli

Scala media   Endolymph fluid

Organ of Corti

Perilymph fluid

Scala tympani

Round window

Basilar membrane

(a)

To the auditory nerve

Scala vestibuli

Bony labyrinth

Perilymph

Scala media

Organ of Corti

Perilymph

Scala tympani

(b)

Tectorial membrane

Sensitive 'hair' cells

Cochlear nerve

(c)

Fig. 76. The mechanism of hearing.

the cochlea. At the point at which the vibrations die out there are nerve fibres conveying this information to the brain.

(e) The brain appears to be able to detect at which point along the cochlea the vibrations fade out. The information detector in the cochlea is the organ of Corti. It is the basilar membrane that vibrates when fluid waves move it. The stimulation of the hair cells causes nerve impulses to pass along the cochlear nerve.

Balance

That part of the ear not concerned with hearing lies within the inner ear. It is the *vestibular apparatus,* made up of semicircular canals, the saccule and the utricule.

The semicircular canals are three in number, lying at right angles to each other in three different planes. These membraneous tubes contain *endolymph fluid;* it is the movement of this fluid, when the head stops and starts motion, that is recorded by the sensory receptors within the tubes.

At one end of each semicircular tube there is a swelling, an *ampulla,* within this swelling there is the sense organ detecting fluid movement. This sense organ is called a *crista*; it consists of hair-like cells which are moved when the endolymph moves. Hair movements create nerve impulses which are sent to the brain.

The *saccule and utricule* contain endolymph like the semicircular canals. Within these structures there are sense organs able to detect movements of the endolymph fluid. The sense organ is called a *macula* and detects changes in the position of the head. An *otolith* structure, a collection of calcium carbonate (chalk) particles stimulate hair-like projections as they are moved about by the fluid motions.

The stimulation of the hair-like sensory structures within the vestibular apparatus causes nerve impulses to flow down the eighth cranial nerve to the brain (cochlear nerve).

THE SKIN AS A SENSE ORGAN

The skin has been considered earlier in some detail when it was seen to be a very large organ.

There are *five* sensations in the skin, but the sensory apparatus connected with all these sensations are not well known.

Pain is probably registered by the diffuse nerve endings in the epidermis and dermis. Tickling is thought to be a stimulation of the nerve endings for pain and touch at the same time.

Touch is registered by sense organs called *Meissner corpuscles,* lying just beneath the epidermis. These are closer together in sensitive areas of the skin such as on the finger tips.

Pressure is sensed by *Pacinian corpuscles* lying deep in the dermis.

Heat and cold is sensed in the skin by nerve endings. There are nerve endings for hot and for cold in the skin as may be demonstrated in the experiment described in the laboratory work part of this book.

THE SENSE OF SMELL

The nerve cells which give man the sensation of smell are located high up within the nasal cavity, towards the front. The nerve cells are contained within the *olfactory epithelium* and are receptors of chemicals dissolved in water. The chemicals are in vapour or gaseous form in the atmosphere and are then taken into the nasal cavity, where they go into solution.

THE SENSE OF TASTE

The organs of taste have been described earlier in the section on nutrition. There are four types of taste bud—sensitive to sweet, sour, salt and bitter—situated on the tongue in different areas.

Sweet and salt is tasted at the tip of the tongue.

Sour is tasted at the edges of the tongue.

Bitter is tasted at the back of the tongue.

The taste buds are sensitive to chemical substances in solution, rather like the sense of smell but at close quarters.

SOME DISEASES AND DISORDERS OF THE SENSE ORGANS

The Eye and Vision

Astigmatism. Poor visual acuity. This is caused by a defect in the curvature of the cornea, the transparent surface through which light passes. The cornea is slightly asymmetrical instead of a part of a perfect sphere. It is corrected by a lens which is also asymmetrical but in the opposite manner.

Blindness. The loss of useful sight very often associated with ageing.

Conjunctivitis (Pink eye). An inflammation of the conjunctiva because of infection by a bacterium or virus.

Colour-blindness. The inability to distinguish between certain colours. The inability to distinguish red from green is the most common defect of this kind and is sex-linked to maleness.

Eye-strain. Working in poor light cannot cause physical damage to the eye but can be tiring.

Glaucoma. An increase of pressure within the aqueous humour which may disturb vision or even cause blindness if not treated. There is an uptake of this fluid into the blood. If for some reason it is not taken back into the blood-stream a high pressure of fluid builds up.

Hypermetropia (Long sight). This is caused by too short an eye or too weak a lens. At rest distant objects are focused behind the retina. Treatment involves the use of convex lenses which bring the light rays into focus upon the retina (see Fig. 74).

Myopia (Short sight). This is caused by too long an eye or too strong a lens. At rest distant objects have their images focused in front of the retina. Treatment requires the use of concave lenses which bring the light rays into focus upon the retina (see Fig. 74).

Presbyopia. This is the normal process of ageing when the lens becomes less able to accommodate because of its loss of elasticity. Treatment involves the supply of glasses for short and distant vision.

Squint. A failure to balance the muscles controlling the movement of the eyes. If untreated one of the eyes which is not pointing in the same direction as the other will be producing images which the growing child will learn to ignore. In effect he uses one eye only.

Stye. A boil on the eyelid.

The Ear and Hearing

Earache. The causes of earache are various and are always worthy of further inquiry. The effects can be very painful and very dangerous.

A boil in the external canal, pus pressing upon the ear drum because of middle ear infection are only two causes of earache in need of attention.

Deafness. Deafness from birth will produce a deaf-mute, one unable to learn speech because of deafness. German measles in early pregnancy has long been known to be a cause of deafness from birth. Deafness can be related to the auditory ossicles, a "deaf aid" can be of help here because vibrations can be lead through the bones of the skull. Nerve deafness (i.e. damage or infection of the eighth cranial nerve) does not permit the use of a hearing aid because the conduction of nerve impulses is the problem not the ear organ itself.

Tinnitus aurium. A ringing, buzzing or roaring sound in the ear because of some ear disorder or associated with nerve deafness.

CHAPTER 12

The Ductless Glands (Endocrine Glands)

A GLAND is an organ that forms a substance and secretes it. Glands are of two types:

Exocrine glands secrete their products into a duct which leads outside the body or into the alimentary canal, e.g. sweat glands, salivary glands, digestive glands.

Endocrine glands secrete their products directly into the blood; they are ductless. The secretion is called a *hormone* and it acts as a "chemical messenger" by circulating in the blood and exerting a profound effect on some other part of the body, e.g. Pituitary gland, thyroid gland, sex glands.

Studying gland function. The knowledge that we have about endocrine glands depends upon the results of the following methods of study.

(a) A study of the gland when it is behaving abnormally, when it is overactive or underactive.

(b) A study of the results when the gland is removed, destroyed or inactivated by chemicals.

(c) A study of the results when the hormone is injected in an experimental animal.

These methods are necessary because we cannot see what a normal gland is doing; its function only becomes obvious when it is *not* doing it!

The chemical nature of hormones. The secretions of endocrine glands are:

169

proteins—pituitary, parathyroid and pancreas hormones,
amino-acids (aromatic)—thyroid, adrenal medulla hormones,
steroids—sex hormones, adrenal cortex hormones.
The endocrine glands of the body to be considered are shown in Fig. 77.

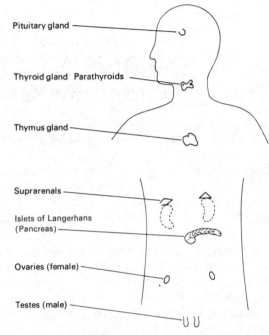

Fig. 77. Endocrine glands and their locations.

THE PITUITARY GLAND

This is a small gland suspended beneath the brain. It is about the
size of a large pea and made up of two parts. The pituitary gland is
often referred to as the *"master gland"* because it produces many
hormones and influences many other glands.

The two parts of the gland have different origins, the front (anterior)
part originating from the tissues of the mouth and nasal cavity, the
rear (posterior) part originating from brain nervous tissue. The two
parts produce different hormones.

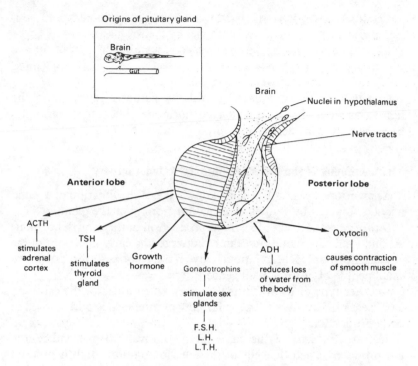

FIG. 78. The pituitary gland.

The anterior pituitary (adenohypophysis) secretes a number of hormones, some of which are listed below. These hormones regulate growth and the activity of many other endocrine glands.

ACTH (Adrenocorticotrophic hormone) This stimulates the adrenal cortex to secrete steroid hormones such as *hydrocortisone (cortisol)*.

TSH (Thyrophic hormone) This stimulates the thyroid gland into activity.

Gonadotrophic hormones. These stimulate the sex glands (gonads) to activity.

FSH (follicle-stimulating hormone) controls the production of ova or sperm. Must be present in order that an individual enters puberty. In older persons it helps to maintain sexual activity.

LH (luteinizing hormone) induces ovulation in females and controls the secretion of sex hormones by the sex glands.

LTH (luteotrophic hormone, prolactin) sustains the corpus luteum during pregnancy and regulates the secretion of milk sometimes known as the lactogenic hormone.

Growth hormone stimulates growth by acting directly upon the tissues, as does the milk-producing hormone. The other hormones have their effect by acting upon other glands.

Malfunctioning of the Anterior Pituitary—Underactivity

As mentioned previously some knowledge of the function of a gland may be obtained by studying the malfunctioning of the gland.

Dwarfism is the result of an underproduction of the growth hormone in childhood. This condition can be corrected by early injections of the missing hormone. These pituitary dwarfs are correctly proportioned small people of normal intelligence.

Another type of dwarfism is produced when areas of the pituitary are diseased. This dwarf type is fat, low in intelligence and stunted in body proportions.

Simmond's disease is the name given to advanced (premature) old age produced when the cells of the anterior pituitary fail to produce secretions.

Malfunctioning of the Anterior Pituitary—Overactivity

Gigantism is produced in children if the growth hormone is overproduced, for some reason, such as a tumour in the gland.

In this case the child grows very tall (as tall as 240 cm) but the body proportions are correct.

Acromegaly is an overgrowth of the bones of the face, hands and feet in adults because of an overproduction of the growth hormone resulting perhaps from a tumour of the anterior lobe. Not only do these bones increase in size but also some internal organs increase in size.

Cushing's syndrome is the name given to a disorder resulting from the overproduction of ACTH. This causes the adrenal cortex to become overactive. The face and body region becomes fat because of upsets in the proper use of food materials, salts and water.

The posterior pituitary (neurohypophysis) is not a true gland. It receives its secretions from the hypothalamus which is close by, and simply acts as a storage organ.

Two separate hormones are found in the posterior pituitary.

ADH (anti-diuretic hormone). This hormone causes water to be retained by the kidneys and so to reduce the loss of water in the urine.

This hormone was known by the name *vasopressin* at one time because when injected into the experimental animal it caused blood-vessels to constrict and so raised the blood pressure.

Oxytocic hormone (oxytocin). This hormone stimulates the contraction of the smooth muscle of the uterus at the end of pregnancy, initiating labour. It also causes milk to be ejected from the mammary glands after child birth. The hormone can be produced synthetically. It has proved of great value in inducing child birth.

An important point to notice here is that these posterior lobe hormones are produced in the hypothalamus which is part of the brain. Here then is a link between the nervous system and the hormonal system.

Malfunctioning of the posterior lobe may reduce the production of ADH (anti-diuretic hormone) and cause a disease called *diabetes insipidus*. In this disorder the urine contains large amounts of water and the person becomes extremely thirsty.

THE THYROID GLAND

This is a gland situated in the neck, consisting of two lobes on either side of the larynx and trachea. It is a soft gland weighing about 25 g lying just beneath the skin. The two lobes are joined making the gland rather like a bow tie. The hormones produced by the gland are iodine compounds which are stored within the gland as a substance called *thyroglobulin*. When needed these hormones are split off from this storage material and released into the blood. The main hormone to be considered is *thyroxine*.

The function of thyroxine is to stimulate metabolism. It increases the combustion of glucose and therefore increases the amount of oxygen used. Body heat is increased. Thyroxine raises the level of body activity, growth and development generally.

Malfunctioning of the Thyroid Gland—Underactivity

Cretinism is produced in children when the thyroid gland fails to develop at birth. Because of the lack of thyroxine the child fails to develop properly, mentally or physically. The child has tough hair and skin, a protruding tongue and is mentally retarded. If left too long this child will fail to develop to normal size or mentality.

Myxoedema is a thyroid-deficiency disorder in adults. The person affected seems usually to be a middle-aged woman, mentally dull, low temperature and low appetite. The skin becomes dry and puffy, the hair becomes thin on the scalp and eyebrows.

Both cretinism and myxoedema can be corrected by the administration of thyroxine orally. (Notice many hormones cannot be taken by mouth as they will be destroyed in the alimentary canal.)

Malfunctioning of the Thyroid Gland—Overactivity

Increased release of thyroxine into the blood-stream causes the metabolic processes of the body to speed up. There is an increased amount of heat produced, the heart beat and breathing rate increases.

Goitre is an enlargement of the thyroid gland; it may be caused by low dietary iodine intake which results in an increased pituitary TSH secretion. The thyroid becomes hyperactive.

Exophthalmic goitre is an overactivity of the thyroid gland accompanied by a bulging of the eyeballs. This more advanced stage of thyroid disorder is also accompanied by symptoms of raised blood pressure, nervousness and excitability.

Overactivity of the thyroid gland is treated by surgical removal of part of the gland or the administration of drugs which reduce the formation of thyroxine in the gland.

THE PARATHYROID GLANDS

There are four parathyroid glands embedded in the surface of the thyroid gland. They are oval in shape and very small, being about 6 mm in length and 20–50 mg in weight. The hormone produced by this gland is *parathormone*.

The main function of this hormone is to regulate the level of calcium in the blood plasma; it does this by acting upon the bone, gut and kidney tubules.

When the level of plasma calcium falls:

Bone—more calcium leaves bone and goes into the blood.

Gut—more calcium taken up from the food.

Kidney—more calcium taken up from the tubular filtrate.

When the level of plasma calcium rises:

A hormone called calcitonin secreted by cells near the thyroid gland, lowers the calcium level in the plasma. This is a hormone which possibly reduces the movement of bone calcium into the blood-stream.

THE ADRENAL GLANDS (SUPRARENAL GLANDS)

There are two adrenal glands located above and a little in front of each kidney. Each gland has two parts to it, an outer *cortex* and an inner *medulla*.

(a) *The adrenal cortex* secretes steroid hormones which are transported in the blood to all parts of the body. For short these hormones are often referred to as *corticoids (cortical steroids)*.

There are three groups of cortical hormones.

Mineralocorticoids regulate the mineral salts in the body. *Aldosterone* is a hormone which acts on the kidney tubules causing them to retain sodium, and to get rid of potassium.

Glucocorticoids regulate the metabolism of carbohydrates, proteins and fats. *Cortisol* (hydrocortisone) is a hormone which promotes formation of sugar from protein thus increasing blood sugar and stored glycogen. Cortisol also has anti-inflammatory actions.

Sex hormones are produced in the adrenal cortex but are not so important as those produced in the sex glands.

Malfunctioning of the Adrenal Cortex—Underactivity

Decreased production of hormones from the cortex can produce *Addison's disease*. In this disorder insufficient sodium is retained by the body and it is lost in the urine.

Muscular weakness and wasting, pigmentation of the skin and intestinal upsets are some symptoms. Treated by administration of adrenal hormones.

Malfunctioning of the Adrenal Cortex—Overactivity

The overproduction of corticoids produces disorders related to the three groups of hormones mentioned previously.

Oedema, the presence of excessive tissue fluid giving a puffiness to the skin.

Cushing's syndrome oedema and obesity, particularly of the face.

Infant Hercules is the name given to a child receiving an excess of male sex hormones from the cortex. The boy becomes precociously developed in his secondary sexual characteristics. His muscles are also well developed.

Virilism in girls and women is the development of male characteristics because of the excess production of male hormones by the adrenal cortex.

(b) *The adrenal medulla* secretes hormones called *adrenaline* and *noradrenaline*. The medulla is under the control of the sympathetic nervous system.

The stimulation of the adrenal medulla causes it to release adrenaline into the blood-stream. This produces the "fight and flight" reactions of the sympathetic nervous system.

Some of the actions of adrenaline are as follows:

Stimulates metabolism.

Increases heart rate, blood pressure raised.

Dilates the pupil of the eye to allow more light to pass.

Dilates blood vessels to skeletal muscles.

Stimulates respiration.

Mobilizes muscle and liver glycogen to increase the amount of blood sugar available.

Inhibits peristalsis of the gut.

The net result of these actions is the tension or fear that accompanies emergency situations.

THE PANCREAS

The pancreas is closely associated in position with the duodenum and stomach. It is a broad band of glandular tissue in the curve of the duodenum. It has two glandular areas, an exocrine digestive area and an endocrine area.

The exocrine area which makes up most of the pancreas secretes pancreatic juice which pours into the duodenum through the pancreatic duct.

The endocrine area which makes up about 1–2 % of the gland secretes two important hormones, *insulin* and *glucagon*. These hormones are secreted from areas of cells or islets of cells described by Langerhans in 1869. It was not until the 1920s that insulin was extracted from the *islets of Langerhans* by Banting and Best.

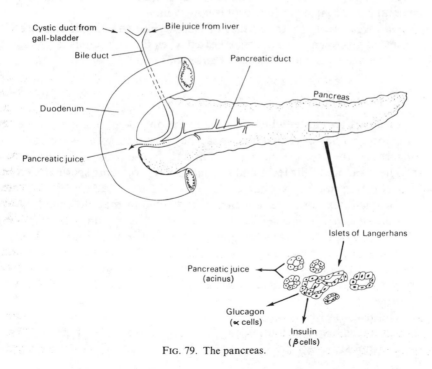

FIG. 79. The pancreas.

Insulin is a protein made up of fifty-one amino-acids in two linked chains. It has now been synthesized in the laboratory.

Insulin is secreted by the β (beta) cells of the islets of Langerhans; its function being *to increase the uptake of glucose* by those cells requiring it. When the blood sugar level rises the secretion of insulin is stimulated which then brings down the level of sugar in the blood.

In *diabetes mellitus* (sugar diabetes) insulin is produced in insufficient quantities so the blood sugar level rises and sugar "spills over" into the urine. The person cannot use glucose and so it passes away in much urine. The person feels thirsty because of the increased urine water loss. Fat is burnt instead of glucose so the person loses weight, but this burning is incomplete and so poisonous *ketones* are produced; this causes the *coma*. Insulin is given by injection usually.

If too much insulin is produced or too much insulin is injected then the blood sugar level is lowered too much. This also leads to coma, and eventual death, if sugar (glucose) is not given.

Glucagon promotes the release of glucose into the blood. Glucose, stored as glycogen in the liver, is released when this hormone is secreted by the alpha cells within islets of Langerhans.

THE TESTES

There are two testes which descend to the outside of the body where it is cooler. They have two functions; they form the sex cells (spermatozoa) and they secrete the male sex hormones (androgens).

The structure of the testis and its function in producing spermatozoa is considered in the next chapter.

The male sex hormone *testosterone* is formed in the cells called the *interstitial cells*. These cells are stimulated by a pituitary hormone described earlier as a gonadotrophin, LH (luteinizing hormone). This pituitary hormone which activates the testis is often called the *interstitial cell stimulating hormone* (ICSH). Once activated the testis hormone produces the *male secondary sexual characteristics*. These characteristics are:

The growth of facial hair.

The male hair distribution of the chest, trunk, pubic region and
    armpits.

The development of a typical muscular physique.

The development of the larynx leading to a deeper voice.

The reproductive organs develop.

THE OVARIES

There are two ovaries situated within the abdomen.

They have two functions; they form the sex cells, the eggs *(ova)* and they secrete the female sex hormones, *oestrogen* and *progesterone*.

The structure of the ovary and its function in producing ova is considered in the next chapter.

The ovary is stimulated into activity at puberty (about 10–14 years) by the sex gland-stimulating hormones of the anterior pituitary gland (gonadotrophins).

The ovary begins to produce mature egg follicles when it responds to the hormone FSH (follicle-stimulating hormone). It is this mature follicle that produces the female hormone *oestrogen*.

A mature follicle releases its egg monthly and the empty follicle converts to a *corpus luteum*. This conversion is stimulated by the LH (luteinizing hormone) from the pituitary gland.

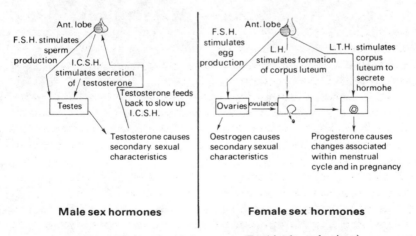

Male sex hormones

Female sex hormones

Fɪɢ. 80. Male and female hormones (Feed-back mechanisms).

From the corpus luteum a hormone *progesterone* is released.

The hormones oestrogen and progesterone bring about the female *secondary sexual* characteristics. These characteristics are:

The development of breasts.

The female distribution of hair in the pubic region and armpits but not present on the face.

The typical female physique with broader pelvis.

The development of the female reproductive organs and the commencement of the *menstrual cycle*.

From this brief introduction to the more important endocrine glands it is possible to see how closely each gland is related to another in function. Glandular function shows delicate interaction.

Our mental and physical attributes are largely determined by our glands. A change in gland function can change our mental abilities, personality or physique.

CHAPTER 13

Reproduction

SEXUAL reproduction involves the fusion of the sex cells *(gametes)*.
Sex cells are produced in the sex organs *(gonads)*.

The male gametes are introduced into the female body where fertilization takes place. The resulting embryo develops within the female to produce a baby.

Gonad ⟶ *Gametes*

Testes (♂) Spermatozoa
→ Fertilization → zygote → embryo → foetus
Ovaries (♀) Ova

THE MALE REPRODUCTIVE ORGANS

The primary sex organs are a pair of *testes* suspended at the base of the abdomen within the *scrotal sac*.

The secondary sex organs are the following:

> Epididymis (pair)
> Vas deferens (pair)
> Seminal vesicles (pair) } internal
> Prostate gland
> The penis

The testes are the male gonads. They have two functions:
  to produce the male gametes, spermatozoa,
  to produce the male sex hormones, the principal one of which is testosterone.

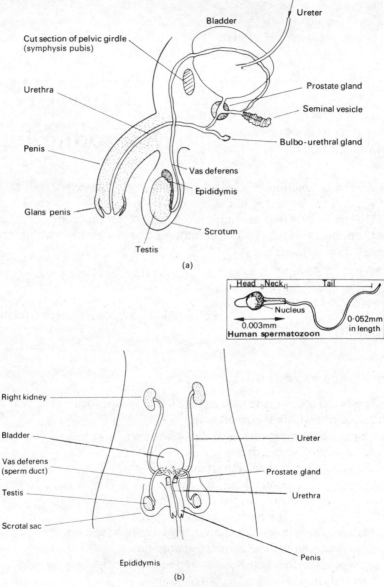

FIG. 81. Male genitals.

The function of the testes as endocrine glands has been considered in the previous chapter. Here we consider the production of spermatozoa.

Spermatogenesis, the production of sperm takes place at a temperature lower than that of the interior of the abdomen. The testes descend from their position within the abdomen before birth. Sometimes one or both of these organs may fail to descend with the consequence that live sperms are not produced because of the higher temperature. Hormone production is quite normal.

Spermatozoa are produced within the many coiled structures called *seminiferous tubules* from puberty to old age. Each spermatozoon is about 0·05 mm long and many millions are produced continuously. If the sperms are not used in sexual intercourse they may be reabsorbed or ejaculated during the night.

The sperms in the seminiferous tubules are not able to move on their own; they are propelled by cilia into the *epididymis* where they then become capable of self propulsion by means of their lashing tails. The sperms are stored in the epididymis before passing along the sperm duct or *vas deferens.* The vas deferens passes up through the groin, *inguinal canal,* into the abdomen.

In the abdomen the seminal ducts pass behind the bladder and then join the urethra. Before the sperms are expelled from the body a seminal fluid *(semen)* is secreted from three structures, the sperm live in this fluid.

Seminal vesicles—produce a viscous fluid which keep the spermatozoa alive and active.

Prostate gland—produces a thinner lubricating fluid.

Bulbo-urethral gland—produce another thin lubricating fluid.

A single ejaculation of semen, about 3 cm³, may contain several hundred million spermatozoa.

Ejaculation of seminal fluid is largely a reflex action. Smooth (involuntary) muscle pushes spermatozoa into the urethra. The involuntary and voluntary muscles of the penis contract to expel the seminal fluid during times of sexual climax.

The penis is the male organ used for passing urine from the bladder to the outside, and for transferring spermatozoa from the testes into the female. It is the organ of *copulation (coitus).* The penis is made up of layers of erectile tissue. This tissue becomes filled with blood at times of sexual excitement and the penis increases in size and length. It becomes erect. The tip of the penis or *glans* is rounded in shape and is covered by a fold of skin. This skin is frequently removed surgically for purposes of hygiene (circumcision).

THE FEMALE REPRODUCTIVE ORGANS

The primary sex organs are a pair of ovaries situated in the pelvic region on either side of the uterus.

The secondary sex organs are the following:

Fallopian tubes (pair)
Uterus
Vagina } internal
Mammary glands (pair)

The ovaries are the female gonads. They have two functions:

to produce the female gametes, *ova,*

to produce the female sex hormones.

The endocrine functions of the ovary have been considered in the previous chapter.

Oogenesis, the production of eggs (ova), takes place in the ovaries from puberty (about 10–14 years) onwards for about thirty years. The time when egg production ceases is called *menopause.* After puberty the ovaries produce one mature ovum each month (until menopause). These ova develop within *Graafian follicles* and are released on about the fourteenth day of the monthly cycle *(ovulation),* the ovum passes into the Fallopian tube where it may become fertilized.

The ovaries are quite small almond-size organs situated near the openings of the Fallopian tubes or oviducts.

The Fallopian tube or oviduct conducts the ovum to the uterus by muscular and ciliary action. It is within this tube that the egg may meet the spermatozoa which swim up there from the uterus. Fertilization then takes place in the Fallopian tube. The fertilized egg (zygote) begins to develop into an embryo even before it has reached the womb (uterus).

The uterus lies in the pelvic region in front of the rectum and behind the bladder. It is about the size of a pear. It is a thick muscular walled hollow organ lined by an *endometrium.* It is this lining which is shed

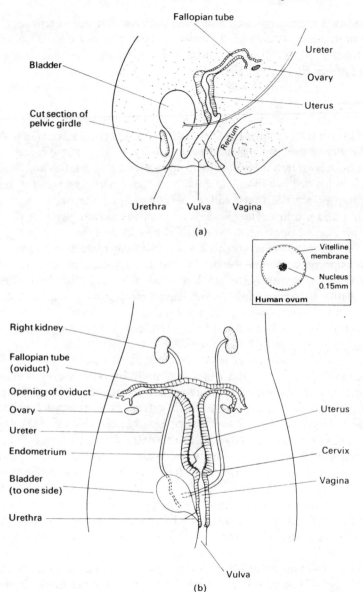

Fallopian tube

Ureter

Bladder

Ovary

Uterus

Cut section of
pelvic girdle

Rectum

Urethra   Vulva   Vagina

(a)

Vitelline
membrane

Nucleus
0.15mm

Human ovum

Right kidney

Fallopian tube
(oviduct)

Opening of oviduct

Ovary

Uterus

Ureter

Endometrium

Cervix

Bladder
(to one side)

Vagina

Urethra

Vulva

(b)

FIG. 82. Female genitals.

monthly if pregnancy does not materialize. A new lining grows.

The uterus has the ability to change, for its weight of 30 g (1 oz) can increase to 1 kg (2 lb) during pregnancy. It returns to more or less its original size and weight after childbirth.

The lower, narrower part of the uterus is called the neck or *cervix;* is opens into the vagina.

The vagina is a muscular passage which receives the male organ during coitus. It is through this passage that the child enters the world.

Externally the female genitalia are called *the vulva*. These consist of a pair of outer larger folds of skin enclosing a pair of smaller inner folds. These folds or lips are called the *labia*. They partially cover the small swelling of erectile tissue called *the clitoris* and the openings of the vagina and urethra. The *hymen* is a small membrane partially closing up the vaginal opening. It is ruptured by physical or sexual activity.

The mammary glands (breasts) develop during puberty because of the increasing oestrogen in the blood. Fat is deposited beneath the skin in differing quantities giving different sized breasts. The size of breast is not related to its value as a milk supplying organ. A small breast can supply milk as efficiently as a large breast.

THE MENSTRUAL CYCLE

A series of changes occur in the female sex organs every cycle of approximately 28 days. For convenience we regard the beginning of this cycle as the time when blood (menstrual flow) is passed from the vagina. The amount of blood lost is usually about 60 ml. This blood loss takes place over 3–5 days of the menstrual cycle.

Menstruation commences at puberty under "instructions" received from the anterior pituitary gland. It finishes at menopause anywhere between 45–55 years of age. At this time the ovaries cease to respond to the anterior pituitary hormones (the gonadotrophins).

The events taking place during the menstrual cycle are as follows:

(a) *Menstrual flow*—approximately 5 days. During this time the uterine lining, the *endometrium* degenerates and is shed. The breakdown of this lining causes some bleeding. The ovum released into the Fallopian tube and as yet unfertilized also passes out of the body in the menstrual blood flow. If the egg were fertilized it would have

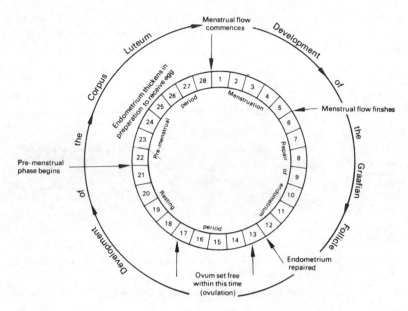

Menstrual flow commences

Development of the Graafian Follicle

Corpus Luteum

Endometrium thickens in preparation to receive egg

Pre-menstrual phase begins

Development of the

Menstrual flow finshes

Endometrium repaired

Ovum set free within this time (ovulation)

period

Menstruation

Repair of endometrium

Resting

period

Pre-menstrual

FIG. 83. The menstrual cycle.

"planted" itself into the thickened uterine lining and the breakdown of this lining would not have taken place.

(b) *Repair and regrowth* of the endometrium continues from about the fifth day of the cycle. The endometrium is being prepared to accept a fertilized egg. After the fourteenth day of the cycle progesterone appears and aids the regrowth of the lining.

(c) *Ovulation* occurs on about the fourteenth day of the cycle. A slight rise in body temperature is seen to occur just after ovulation and it continues until the end of the cycle.

The ovum is released from one or other of the ovaries and is sometimes accompanied by pain. The egg passes into the Fallopian tube where it may be fertilized by a sperm if sexual intercourse has taken place. The egg is moved down to the uterus by ciliary and muscular action in the Fallopian tube. The egg remains in the uterus for the next fourteen days or so, until the next cycle begins, if it has not been fertilized.

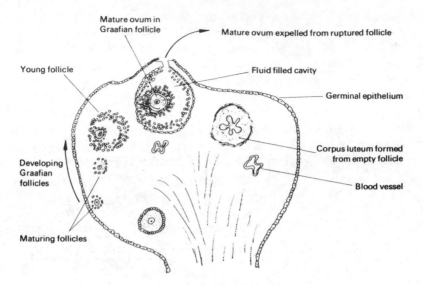

Mature ovum in Graafian follicle

Mature ovum expelled from ruptured follicle

Young follicle

Fluid filled cavity

Germinal epithelium

Corpus luteum formed from empty follicle

Developing Graafian follicles

Blood vessel

Maturing follicles

FIG. 84. The Graafian follicle in the ovary.

The ovum is released from the Graafian follicle. The empty follicle becomes a glandular structure called the *corpus luteum*. It is from the corpus luteum that the hormone *progesterone* is secreted. This hormone prevents further eggs from being released.

The corpus luteum and its secretion persists until the end of the cycle if no egg is fertilized. If the egg is fertilized the corpus luteum persists and the changes associated with pregnancy result.

The menstrual cycle halts during pregnancy. If no fertilization has taken place then the menstrual cycle recommences after a period of menstrual flow which removes the endometrial lining.

FERTILIZATION AND PREGNANCY

Sexual intercourse (coitus) normally results in the fertilization of an ovum released into the Fallopian tube at ovulation. There are periods during the menstrual cycle when fertilization does not take place or is less likely to take place. This is often referred to as the "safe-period".

It is that part of the cycle away from the ovulation period. The ovum survives only for a few days in the Fallopian tube and so coitus taking place when the egg has degenerated cannot result in pregnancy. This "safe-period" is used by some as a basis of *birth-control,* but cannot be considered as really "safe". The time of ovulation can vary from time to time and so pregnancy can result.

(a) *Fertilization* results when a living sperm meets a living egg. The sperm penetrates the outer coat of the egg and moves in towards the egg nucleus. The nuclei of sperm and egg unite. The result is called a *zygote.* It is this fertilized ovum that develops into a baby.

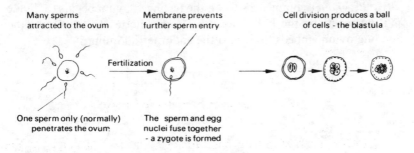

FIG. 85. Fertilization and cellular division.

The sperms are brought into contact with the egg by means of the sex act (coitus). The introduction of spermatozoa to the femal sex organs as it passes along the Fallopian tube to the uterus. In the uterus the dividing ovum embeds itself into the endometrial lining.

Two people engaging in "love-play" bring about changes in the sex organs. The male penis becomes erect for reasons mentioned previously. The female vulva becomes swollen, the clitoris becomes erect and mucus secretions lubricate the vagina and vulva.

The male organ is introduced into the lubricated female opening where friction takes place between the glans penis and the vaginal walls. These frictional movements on the organ heighten sexual excitement in both male and female. Breathing rate increases, pulse rate increases and blood pressure rises. Both partners come to a point of excitement called *orgasm* when the male ejaculates semen into the vagina. The orgasm may not occur at the same time in each partner and requires experience and understanding for it to do so. The male will normally experience one orgasm whereas the female may have more than one during a sex act.

The sex act so described appears simple and devoid of feeling. This act is open to much moral discussion and much mental anguish for some. The sex act can be associated with guilt because of some earlier training and the inability to perform correctly is known as *impotence*. *Infertility* is the inability of the sex organs to produce a successful fertilization. The latter may be open to medical treatment, the former to psychological treatment.

(b) *Pregnancy* commences when the fertilized ovum begins to divide as it passes along the Fallopian tube to the uterus. In the uterus the dividing ovum embeds itself into the endometrial lining.

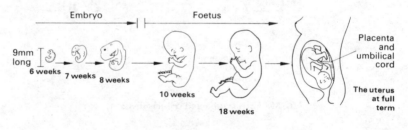

FIG. 86. Embryonic development.

The usual sign of pregnancy is the absence of menstrual bleeding at the expected time. This is not always true because the menstrual cycle and its regularity may be upset by a change in life circumstances.

Normally one egg is released from the ovary at ovulation but if two ova are released and fertilized *twins* are the result. This release of more than one egg may be a family characteristic or it can result when a woman takes a fertility drug to induce ovulation.

Twins are of different types:

Fraternal twins are those produced when *two eggs* are fertilized on one occasion. They can be of different sexes and as different as any other two children in the family. They just happen to have the same birthday.

Fig. 87 An Embryo in the 4th week

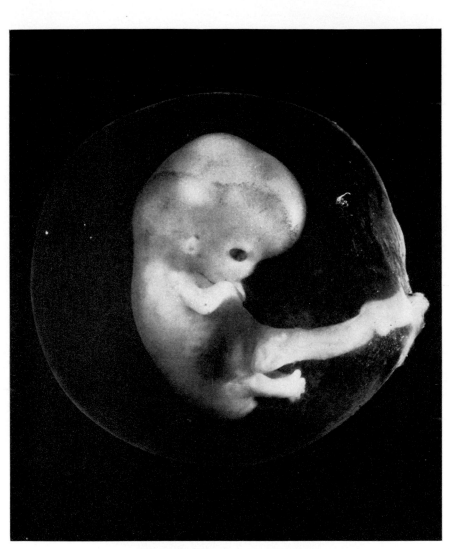

Fig. 88 An Embryo in the 7th week

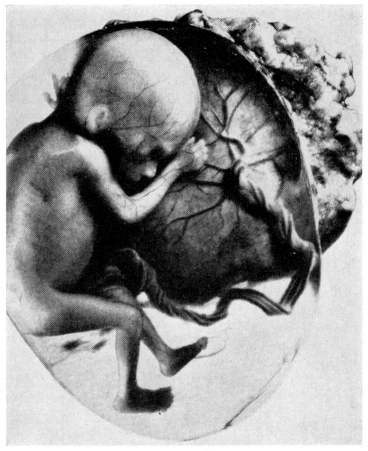

Fig. 89 Foetus at 4 months

Identical twins are produced from *one fertilized egg* which splits into halves, each of which develops into an embryo. These children are identical genetically and in all their main physiological characteristics.

Pregnancy changes are brought on by the progesterone hormone released from the corpus luteum, and later, from the placenta. These changes are designed to nourish and protect the developing baby.

Pregnancy may be calculated as the time between the cessation of the menstrual period and the birth of the child. The day or week of conception is not easy to determine and so for convenience one calculates from the first day of the last menstrual period; this gives a pregnancy or *gestation* period of 40 weeks.

Pregnancy may be diagnosed by various tests because of hormones being found in the urine about a month after conception. For instance, the clawed South African toad (called *Xenopus*) kept in many biology laboratories may lay eggs if injected with a sample of pregnant woman's urine. Mice and rabbits are sometimes used for similar tests. These methods are not used for regular pregnancy testing nowadays.

Modern pregnancy tests are the *agglutination inhibition* tests which use the knowledge that gonadotrophins in pregnancy urine inhibit the agglutination of added test materials. This is usually effective 6 weeks after the last period.

During pregnancy the woman is likely to eat more food but more particularly she is liable to choose foods which are of value to the developing child. Iron- and vitamin-containing foods are particularly important. The *overdosing* with vitamin concentrate tablets during pregnancy or at other times is thought to be harmful.

The period of pregnancy is treated by some as a time of undue delicacy but the female is in fact a very robust organism and thoughts of pregnancy as some sort of illness are rather old fashioned. Contact with diseases, particularly German measles, smoking and over indulgence in alcohol or drugs are the main areas of concern for the pregnant woman.

CHILDBIRTH (PARTURITION)

The developing child is attached to the mother by way of the *placenta*. The child is attached to the placenta by the *umbilical cord*.

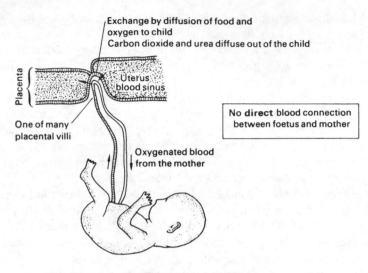

Exchange by diffusion of food and oxygen to child
Carbon dioxide and urea diffuse out of the child

Placenta

Uterus blood sinus

No **direct** blood connection between foetus and mother

One of many placental villi

Oxygenated blood from the mother

FIG. 90. The placenta.

The placenta is firmly attached to the uterine lining and it persists throughout pregnancy in order to perform the following functions:

   (i) It secretes hormones, oestrogen, progesterone and a gonado-trophic hormone. It is the latter which appears in the urine of the pregnant woman.

  (ii) It conducts the nutrients and oxygen from the mothers blood, by diffusion, into the blood stream of the developing child. The two blood-streams are not continuous one with another.

 (iii) It conducts carbon dioxide and other waste products from the child to the mother, again by diffusion from one blood to another.

At the birth the placenta and child are expelled from the uterus by muscular contractions called *labour*. The stages of labour may be described as follows in the first birth. (In other births these times may be shortened.)

   (i) *The first stage of labour* puts a pressure on the fluid surrounding the child. Muscular contractions of the upper part of the uterus increase in strength and frequency lasting for about 15 hours.

The membranes in advance of the child's head burst and some fluid is released.

(ii) *The second stage of labour* may last up to 2 hours when the muscular contractions of the uterus increase and the child begins to move downwards, head first. The mother helps this process of birth by "pushing" with the abdominal muscles.

(iii) *The third stage of labour* occurs up to a quarter of an hour after the birth of the child. Some further contractions of the uterus push out the placenta and membranes. This material is known as the *afterbirth*.

The umbilical cord is ligatured (tied off) and later cut.

There are occasions when it is necessary to *induce labour*. The mother may be given warm abdominal baths followed by injections of a pituitary hormone (oxytocic hormone) which increases muscular contractions of the womb.

The Problems of the New-born Baby

The unborn child grows in a constant environment within the mother. It has a surrounding fluid at body temperature (37 °C; 98·6°F); it receives a constant supply of oxygen from the mother, food nutrients are also supplied. At the birth the new baby must experience the following changes:

(i) It becomes exposed to air with its usual temperature changes. The baby must now regulate its own temperature. The new-born baby has special stores of fat, called "brown body fat" which is capable of rapid usage in order to supply heat. Premature births present a problem because this fat and other fat is in short supply.

(ii) The foetus receives its oxygen supply from the mother by the way of her blood flowing within the placenta. The gas diffuses across from the mother to the developing baby. The foetus does not use its lungs and so the heart does not pump blood to the lungs. At birth the lungs come into action and the heart must now pump blood to them.

In the uterus the foetus has been living in a lower concentration of oxygen and so the blood pigment haemoglobin is designed for this type of life. At birth the haemoglobin has a higher concentration of oxygen brought to it in the air taken in the lungs. At birth the child gasps when entering the colder air bringing in enough oxygen to start the life giving automatic breathing mechanisms.

(iii) Metabolic disorders of the new born are rare but it is important that they be detected.

Phenylketonuria is an inborn metabolic disorder which appears just after birth. The baby cannot use a chemical phenylalanine and it therefore accumulates in the blood. It is poisonous to the brain and leads to mental deficiency. It can be tested for in the urine of the baby, these babies must be fed on low regulated amounts of phenylalanine.

CONTRACEPTION

One method of preventing conception has been mentioned, that is confining sexual intercourse to the time of the menstrual cycle described as the "safe-period". This is not so very "safe" as already mentioned.

Birth control is regarded by many as a most important method of regulating the growth of human populations. Where populations grow to such an extent that they are too numerous to feed and cloth properly, a control of their numbers is necessary in order to prevent disease and suffering.

Some methods of contraception provided by modern society are mentioned below.

(i) *Coitus interruptus* is a popular but unsatisfactory method of preventing fertilization. The male withdraws from the female at such a time as to prevent sperms entering the vagina.

(ii) *Physical barriers* to sperm have been very popular. These devices prevent sperm from coming into contact with the egg. Rubber devices worn by the male or devices inserted within the female prevent the sperm from reaching the egg. These devices are not always satisfactory. Intra-uterine devices inserted within the uterus aim to prevent the fertilized egg from implanting itself there.

(iii) *Chemical contraceptives* are agents applied to the female vagina in order to kill sperms or immobilize them.

(iv) *Hormone preparations* are the most reliable agents to prevent fertilization of the egg by the sperm. They prevent the release of an egg from the ovary. These oral preparations are included within "the pill". An oral contraceptive pill for men is under study at the present.

(v) A final solution to repeated, unwanted pregnancies is *sterilization*. This involves a surgical operation. In men the operation is a minor one which involves cutting the tube leading sperm from the testes to the exterior (vasectomy).

In women the oviducts are cut.

Contraception and abortion (ending pregnancy) are topics attracting much discussion because they are issues linked strongly with moral and religious values.

SOME DISEASES AND DISORDERS OF THE REPRODUCTIVE SYSTEM

Amenorrhoea. The absence of menstrual periods amongst girls and women of child-bearing age. It may be a symptom of some emotional upset or, most obviously, pregnancy.

Breech delivery. Normally a baby enters the world head first. A breech delivery is where the baby has his backside first. He is usually turned to the head-down position during confinement, where possible.

Caesarian section. This is a surgical operation during the late stages of pregnancy. The baby is removed through the opened up abdominal wall and uterus. This is done when normal birth is not possible or advisable.

Cervical cancer. Cancer of the neck of the womb (the cervix) may be detected in its early stages by cervical screening tests where a few cells are examined from this area. The disease can be cured if discovered early.

Impotence. The inability of a couple to perform sexual intercourse. This is more common in men because of failure to secure an adequate erection of the penis. In the main the reasons are emotional. Worry about the sexual part of life is one possible cause. Female impotence may also be caused by an anxiety state (frigidity).

Infertility. The inability of a sperm to fertilize an egg. The reasons could be amongst the following: inactive or insufficient spermatozoa; eggs failing to come into contact with sperms; eggs not being released from the ovary.

Venereal disease (VD). Infectious disease transmitted only by sexual activities. This type of disease can cause unpleasant and damaging results if not treated early. All these diseases seem to respond to treatment with drugs such as antibiotics if detected early. Two venereal diseases are:

Gonorrhoea—a bacterial infection of the mucous membranes of urethra and uterus. After a short time pain and pus is discharged, less obviously in the female. Unchecked this disease can cause sterility and disease of the joints and eyes.

Syphilis—a bacterial infection (a spirochaete) passed on by sexual contact. This spirochaete can infect any unborn child by passing across the placenta. The effects of this infection may commence as soft sores in the region of the genitals. If not treated the disease may progress to the most horrible stages with possible death through paralysis and insanity (GPI—general paralysis of the insane).

CHAPTER 14

Genetics

GENETICS is the study of heredity. What we inherit depends upon our *genes,* which we receive from our mothers and fathers. Genes are assembled within the nuclei of all cells as structures called *chromosomes.*

We inherit not only genes but also the cytoplasm contents of whole sex cells. The cytoplasmic contents are synthesized upon the basis of information supplied by the genes.

GENES AND CHROMOSOMES

Genes are made up of special acids found in the nucleus, the *nucleic acids.* The main nucleic acid in the nucleus is *deoxyribonucleic acid (DNA).*

DNA is a very special substance because it is this which controls the production of all the proteins in the body. So, one unit of DNA governs the synthesis of one protein, usually an enzyme. There are many thousands of proteins making up the body, so this means DNA must have some special abilities.

The structure of DNA molecules will reveal that there are many different types and so they can govern the production of many different proteins. The way in which the DNA molecules differ one from another can be seen in the diagram overleaf.

The structure of a DNA molecule has been shown to be *helical* (a double helix like a spiral staircase). The names of Watson and Crick of Cambridge University are associated with the work which led to the discovery of this structure.

There is another nucleic acid, found mostly in the cytoplasm, called

197

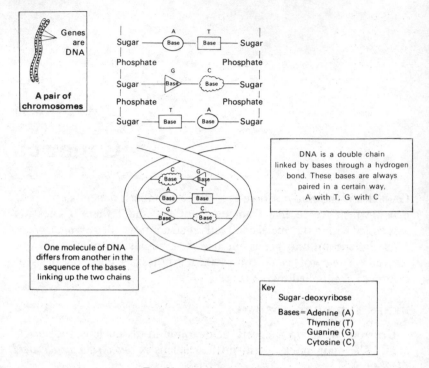

FIG. 91. A molecule of DNA.

ribonucleic acid (RNA). This acid is in some way made under the instructions given by the various units of DNA in the nucleus. So, for each unit of DNA a unit of RNA is produced. This moves out of the cell nucleus into the cytoplasm. These units of RNA in the cytoplasm "collect up" amino-acids and link them together to make a protein. This building activity takes place on ribosomes.

We inherit the information units DNA from our parents. We are able to inherit this DNA because when a cell in a sex organ divides (meiosis) to produce a sex cell the DNA also divides in half. The DNA spiral is able to unwind, one-half passing into each cell. It can later reproduce the missing half.

As mentioned previously, DNA is organized into structures called chromosomes.

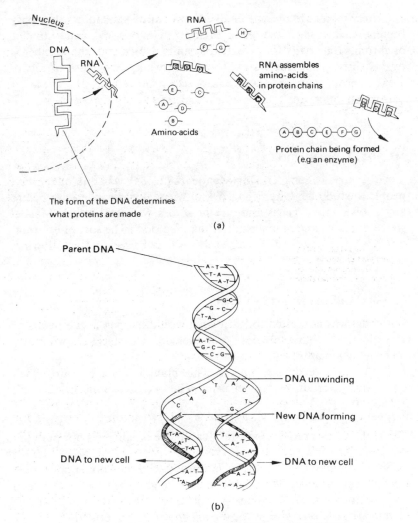

FIG. 92. (a) Proteins being made. (b) DNA at cell division.

During meiosis chromosomes "cross over" and exchange DNA, thus producing the variation of an organism from one generation to the next (see p. 9).

Chromosomes are present in all cells as paired structures; they only become visible when the cell is dividing. There is a definite number of chromosome pairs for a particular animal. For man the number is twenty-three pairs (forty-six). There are twenty-two pairs of *(autosomes)* chromosomes plus one pair of sex chromosomes. It is the latter pair which determine maleness or femaleness.

MENDELIAN INHERITANCE

An Austrian monk Gregor Mendel (1822–84) was one of the first people to study heredity. The work of Mendel was carried out before any knowledge of genes and chromosomes was established. Mendel spoke of plants and animals inheriting "factors". The sort of information given to us by Mendel's experiments may be seen in the following.

Crossing Tall and Dwarf Pea Plants

Mendel was interested to learn what would happen if two plants of very different characteristics were crossed. He selected two pure-breeding pea plants:

(i) A pure-breeding *tall plant*. This plant is a pure-strain which always produces tall offspring when crossed with another tall plant.

(ii) A pure-breeding *dwarf plant*.

The result of this crossing is a tall plant. The offspring is not pure tall but a mixture or *hybrid*.

From this crossing it is clear that no dilution has taken place producing a plant which is an intermediate height. One factor (tallness) has dominated over the other parent factor (dwarfness). The dwarf characteristic is *recessive* to the *dominant* tall characteristic.

The experiment is continued to find out what type of pea plant is produced when two hybrids are crossed. This crossing shows that the dwarfness factor has not been lost in the hybrid but can segregate out.

The result of this hybrid crossing is:

Talls (75%)   Dwarfs (25%).

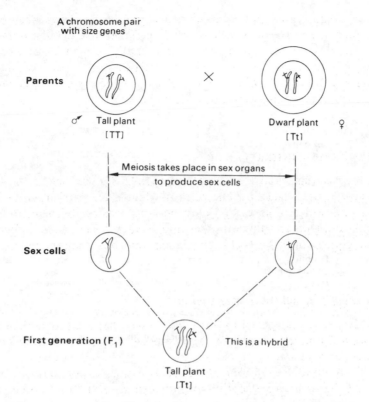

FIG. 93. A dominant and recessive.

From Fig. 94 it can be seen that all the talls are not genetically the same. The talls are TT (pure-breeding), Tt (hybrid tall). In order to distinguish one type of tall from another the following two terms are useful:

Genotype—the "gene type", as indicated by the symbols we use to represent the genes.

Phenotype—the "physical type", as revealed by the physical characteristics such as colour or size. The phenotype may be affected by environmental factors, e.g. food.

The hybrid tall has a genotype Tt and a phenotype tall. The genotype tells us it is not pure.

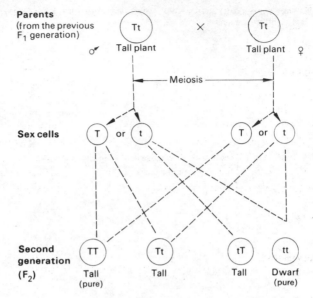

FIG. 94. Crossing hybrids.

The genotype can be described by two terms:
Homozygous—pure-breeding, containing like genes,
    for example TT or tt.
Heterozygous—hybrid, containing unlike genes,
    for example Tt.
Mendel's work tells us about dominance and recessiveness in carefully controlled experiments. These results will not necessarily be seen in nature as the crossings are not so regulated. Similarly, human characteristics are not so simple, tall or short, black or brown. There are many "shades" of size or colour regulated by many genes.

Blue and Brown Eyes

A common illustration of dominant and recessive genes in man is eye colour inheritance.

The brown eye is dominant over the blue eye, as can be seen in the diagram below.

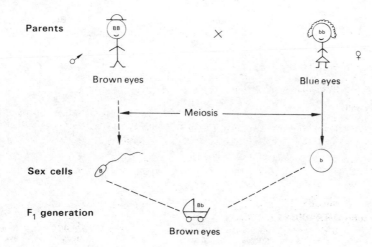

FIG. 95. Eye colour inheritance.

The example in Fig. 95 assumed both parents to be "pure" (homozygous) for the colour concerned; this is not very likely. It is more likely that the parents are "hybrids" for the particular dominant gene. Figure 96 shows that it is possible for two brown-eyed parents to have a blue-eyed child.

EYE AND HAIR COLOUR INHERITANCE

As mentioned previously, human genetics is not really a simple matter of one gene, one character. Many genes may interact to produce one character. Let us now look at a simplified version of the inheritance of two characteristics, that is eye colour *and* hair colour.

What characteristics will be inherited by the children of the following parents?

Father: Red hair and brown eyes.

Mother: Brown hair and blue eyes.

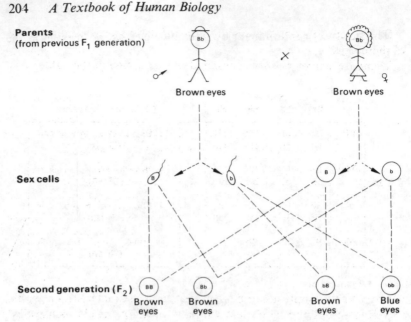

FIG. 96. Crossing eye colour hybrids.

Before attempting genetics problems we need to produce symbols to represent the genes; we will use the following symbols:

B = Brown eyes (dominant character)
b = Blue eyes (recessive character)
H = Brown hair (dominant character)
h = Red hair (recessive character)

The genotypes of the parents may be:

Father: Hair (red)   *Mother:* Hair (brown)
        hh   HH or Hh (more likely)
        Eyes (brown)   Eyes (blue)
        BB or Bb (more likely)   bb

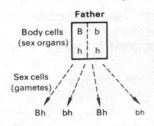

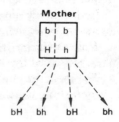

B. b
8 8

a 331

Sexual reproduction will result in one of these sperm fertilizing one of these eggs.

There are sixteen possible combinations as shown in the table:

		Sperm				Child's characteristics
		Bh	bh	Bh	bh	
	bH	Bb Hh	bb Hh	Bb Hh	bb Hh		4 Brown eyes, brown hair (Bb Hh)
Eggs	bh	Bb hh	bb hh	Bb hh	bb hh		4 Brown eyes, red hair (Bb hh)
	bH	Bb Hh	bb Hh	Bb Hh	bb Hh		4 Blue eyes, brown hair (bb Hh)
	bh	Bb hh	bb hh	Bb hh	bb hh		4 Blue eyes, red hair (bb hh)

From this example it can be seen that these characteristics are able to segregate out and be present with any other of the characteristics. In other words, red-haired people can have blue or brown eyes, for example. Red hair is not *linked* with any particular eye colour.

Linkage is an important aspect of human genetics. Some characteristics are linked with a particular sex. For instance, colour blindness is linked to maleness. In order to understand this we need some knowledge of how our sex is determined genetically.

SEX DETERMINATION

In the human body cells there are twenty-three pairs of chromosomes.

In the female there are twenty-three pairs of matching chromosomes but in the male the twenty-third pair do not match.

The female sex chromosomes (twenty-third pair) are similar in size; they are called the X chromosomes.

The male sex chromosomes are not similar in size; one is a larger X chromosome but the other a smaller one called the Y chromosome.

A female cell has twenty-two pairs of autosomes and one pair XX sex chromosomes.

A male cell has twenty-two pairs of autosomes and one pair XY sex chromosomes.

Sex is determined in the manner illustrated in Fig. 97. From this

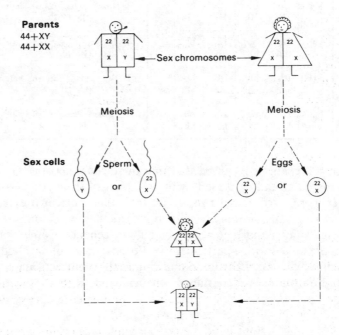

The chances of a boy or of a girl are 50/50

FIG. 97. Sex determination.

diagram we can see that there is an equal chance of a boy or a girl baby being produced.

A boy is produced if an egg is fertilized by a sperm containing a Y chromosome. A girl is produced if an egg is fertilized by a sperm containing an X chromosome.

Sex-linkage

If a gene is on the X chromosome in a male cell then it has no matching chromosome with genes to oppose its action. For example, *a red–green colour* blindness gene on the X chromosome has no gene on a matching X chromosome to cancel it out or dominate it. A male with the recessive red–green colour blindness on the X chromosome will be colour blind.

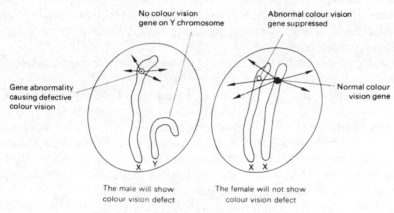

No colour vision
gene on Y chromosome

Abnormal colour vision
gene suppressed

Gene abnormality
causing defective
colour vision

Normal colour
vision gene

X Y

X X

The male will show
colour vision defect

The female will not show
colour vision defect

FIG. 98. Sex-linkage and colour-vision defect.

The female can carry this recessive abnormality but will not show it because she has an X matching chromosome with the normal, dominant, vision gene. The abnormality is suppressed by the normal dominant gene on the matching X chromosome.

A female can be colour blind if a rare circumstance occurs when she inherits an abnormal gene on both X chromosomes.

Haemophilia is another example of a sex-linked recessive character. The female carries this disorder but the male suffers from the disease. The female can only suffer from this bleeding disorder if she has a father who is haemophiliac and a mother who is a carrier. This is a very remote possibility.

There are abnormalities in the sperm or egg which result in odd chromosome patterns, for instance:

Turner's syndrome—chromosomes 44 + XO. Female characteristics on an immature "male" intersex.

Klinefelter's syndrome—chromosomes 44 + XXY. Male characteristics but abnormalities in the testes cause sterility.

SOME GENETIC ABNORMALITIES

Mongolism (Down's syndrome). This is the result of faulty egg production where instead of forty-six chromosomes the fertilized egg results in forty-seven chromosomes. Instead of there being twenty-three pairs of chromosomes, number twenty-one has three instead of two chromosomes and so the result is twenty-three pairs plus one single chromosome.

A mongol child shows mental and physical retardation accompanied by a facial resemblance to the Mongolian races.

Sickle cell anaemia is the result of a *mutation* of the DNA (gene). A mutation is a slight chemical change in the structure of the DNA. This slight change in DNA will produce a change in the protein manufactured. A change in the type of haemoglobin manufactured can make a person anaemic. Such a mutation will not allow a person to survive for very long. Most mutations are disadvantageous to our survival. Atomic radiations can produce gene mutations.

In West Africa where malaria is widespread, people with this sickle cell anaemia appear to be able to survive the malarial disease better. So here a mutation has an advantage and so it is spread throughout large areas of the population. It has survival value.

Sex Abnormalities

Disputes over the sex of an athlete sometimes happen and the test of sex is carried out as follows in the laboratory.

Cells taken from the body (a mouth scraping, for instance) are stained and the nucleus is examined. Females show an area of stained material within the nucleus called a chromatin positive area; males do not show this area. More refined techniques may be used involving a more detailed study of the chromosomes.

Some studies point towards a probable association between criminal forms of behaviour and sex chromosome abnormalities. This has yet to be substantiated as the studies are at an early stage.

CHAPTER 15

Man in Towns

Homo sapiens makes his home in widely differing climates and geographical terrains. The survival of his species under these different conditions is partly due to his ability to insulate himself against his environment.

Man lives in houses in communities. This fact presents problems:

(a) Suitable housing.
(b) Ventilation.
(c) Heating.
(d) Lighting.
(e) Water supply.
(f) Disposal of waste.

THE PROBLEMS OF LIVING IN TOWNS

Man living in houses can create a comfortable environment for himself and his family. He can reduce his worries about keeping the weather out, obtaining fresh water, getting rid of his wastes. But what happens when several million others are doing this? Our environment begins to show varying degrees of pollution.

Overcrowding

Without some degree of control our living space becomes over-crowded. Our natural food supplies begin to get less and so we have to synthesize foods. Many people begin to suffer from inadequate food

THB—P

supplies. Malnutrition and death result. People crowded together transmit their diseases to each other and they "get on each other's nerves". Crowds bring stress. Competition between man and man results in social and physical violence. Unhappiness and bad health can result from overcrowding. What is the answer? Birth control?

Environment and Pollution

People are selfish. The world is large and so many people regard it as a convenient and very large "dustbin". How many times have you thrown papers or packets out of a car, bus or train window? We tend to think it will rot, disappear or somebody will deal with it. Multiply that type of thinking by millions and see what a great deal of litter covers the world's surface.

Another example of this thoughtlessness is the release of poisonous, dirty or offending gases into the atmosphere. A car puffs out poisonous carbon monoxide, a factory chimney puts out soot, tar and sulphur dioxide. A selfish cigarette or pipe smoker pursuing his dirty and death-dealing habit in a cinema or restaurant is an all too common cause of pollution and anger.

It is worth remembering that from early days smoking tobacco has been known to be a dirty and offensive habit. Today we have another reason to stop it: smoking can lead to a painful, suffocating death.

Another example of our selfishness leading to atmospheric pollution is noise. Excessive and repeated noise has been shown to be a cause of stress in many people. Just imagine living near a large airport or in a house with a fanatic "musician" tuning his instrument at all hours as you try to read a book or listen to the radio.

Many forms of pollution are punishable in law, but what a bother it is to go to law when a small act of non-selfishness could correct the situation.

Ecology and Conservation

Waste fuels dumped in river or lake or sea cause death for millions of living organisms. Chemical wastes dumped in ditches or carelessly cast out will poison the soil and eventually our food. Chemicals used to control insect populations or used to encourage growth in food animals are all liable to find their ways into our stomachs.

Because of atmospheric pollution man has had to take diverting action in order to conserve his plant and animal environment. Not only living creatures suffer as a result of pollution but also man-built structures of historical or artistic interest.

Ecology is the study of plants and animals in their natural environment with a view to learning how their lives are interrelated. Pollution can rupture the natural association between plant and animal. A little of man's town environment will now be studied.

UNSUITABLE AND SUITABLE HOUSING

A house may be regarded as *unsuitable* for human habitation if it shows many of the following features.

Unsuitable Housing

Bad site. Built on marshy land, near rivers, factories or sewage outlets. With an unpleasant outlook, too close to other houses, on shifting land, away from main drainage or water supply.

Bad design. Unimaginative use of building materials, small windows facing other buildings, outside doors opening into living quarters. Small kitchen, no bathroom, outside non-flushing toilet. No damp-proof course in walls. Wastage of space. Badly ventilated and illuminated corridors.

Lack of repair. Uneven walls, door and window frames. Falling masonry, leaking walls and roofs, decaying woodwork. Unpainted dirty internal walls.

Bad sanitation. Open sewage pipes in close proximity to drinking water, children and animal life. Inadequate non-flushing toilets used by many people. Unsanitary water supply, inadequate washing facilities with no hot water. Overcrowding. Poor ventilation.

A house may be regarded as *suitable* for human habitation if it shows many of the following features.

Suitable Housing

Site. Dry porous soil and sub-soil above level of flooding rivers, lakes, etc. Pleasant sunny outlook for main rooms. Good sewage, water and electricity supplies. Close to shops and public services. A reasonably private garden.

Design. Imaginative use of building materials, stone, brick, concrete or timber. Planned layout of rooms with adequate use of space. Use of concrete foundations for buildings on soft land. Cavity walls with adequate damp-proof courses. Sound-proof partition walls. Roof design adequate to protect the inhabitants from the prevailing climate. Floors constructed of or covered by materials which are easily cleaned. Windows providing adequate illumination and ventilation.

Sanitation. Adequate, closed sewage pipes, well separated from the drinking water supply. Flushing toilets, hot water supply and bath. Adequate sleeping accommodation to prevent overcrowding. Good ventilation to prevent dry rot and other fungi which decay wood.

VENTILATION

Modern man spends much time in the confines of four walls, and may thus experience much discomfort if his buildings are not adequately ventilated. The cause of this discomfort is the still, moist air which does not allow surface sweat to evaporate and the body to be refreshed. The percentage of water vapour in the air, *the humidity,* may be regulated by adequate ventilation. Ventilation may be achieved in two ways :

(i) Natural ventilation.

(ii) Artificial ventilation.

(i) *Natural ventilation* takes place through windows, doors or special devices fitted to allow the passage of air. Ventilation of this type relies upon the movement of air outside the building.

A certain amount of ventilation takes place by the diffusion of gases, or by convection currents from a heater or fire.

(ii) *Artificial ventilation* relies upon mechanical devices to move the air.

Electric extractor fans placed in windows or walls remove foul air to the outside. An alternative is to propel air into a room, for this will then displace the foul air in the room.

A combination of both the above methods is commonly used.

HEATING

The temperature of the air in contact with man's skin should be maintained at a comfortable optimum (approximately $18°C, 65°F$). The air temperature is maintained near this optimum by means of the following two factors:

(i) Clothing.

(ii) Space heating.

(i) *Clothing* in very hot countries covers the whole body, is loosely fitting and is of light material. The function of clothing here is to protect the skin from the heat rays of the sun, at the same time allowing adequate ventilation to cool the body. In cooler climates clothing is required to keep the body warm by surrounding it with an enclosed blanket of warm air.

(ii) *Space heating* may be either direct or indirect.

Direct heating occurs when the source of heat is in the room that is being warmed. A coal, gas or electric fire and various closed stoves are sources of direct heat.

Indirect heating occurs when the heat is generated at another point and transferred to the room which is to be warmed. Central heating is an example of this type of heating where hot water, steam or hot air are circulated around the building.

LIGHTING

Light is necessary before the material world will become visible to the human eye.

Light may be of two types:

(i) Natural light.

(ii) Artificial light.

(i) *Natural light* comes from the sun during day-time. Sunlight is slightly bactericidal and is an agent in the production of vitamin D in the skin. Daylight enters by way of windows and either comes directly from the sky, or is reflected from internal or external surfaces.

(ii) *Artificial lighting* is produced by gas or electric devices which emit light rays. The rays which reach the illuminated object may be:

(a) Direct—straight from the source or a reflection from the source.
(b) Indirect—reflected from walls and ceilings.
(c) Diffused—coming from all directions.

The optimum illuminations necessary in order to carry out tasks of various kinds without discomfort have been tabulated by lighting engineers.

WATER SUPPLY

Men in communities normally live near some source of water: a river, a lake or a spring. The source of this water is rainfall. Before it is drunk, water must be purified of some organic and inorganic material.

The organic material may be the pathogenic organisms of typhoid, cholera, dysentery or some larval stage of a parasitic worm or fluke.

The inorganic material may be iron and lead salts, sand and silt, or salts of calcium causing hardness of the water. The most important factors that must be removed from the water before human consumption are the organic materials.

The purification of water takes place as follows: *sedimentation* is the first procure. The water is collected in reservoirs and all heavier, solid particles settle to the bottom. During this storage period of about 30 days a large percentage of bacteria are destroyed.

Reservoirs must not be contaminated from outside sources such as sewers or animals.

After sedimentation is *filtration*. This involves passing the water from the reservoir onto sand filter beds. The filters may be slow or rapid sand filters.

The slow sand filter is a few feet in depth and consists of a layer of sand on top of a layer of gravel, graded from fine to coarse. The water percolates down through these layers. After 3 days the sand layer is covered with a film of green algae, bacteria and other organic material. This film is important in the purification process.

The rapid sand filter may function using pressure or the force of gravity. Chemical coagulants are added to the water in order to speed up the formation of an organic film across the top layer of sand. This film will form in a few minutes, compared with the days needed to create a biological film.

Sterilization completely destroys any remaining pathogenic organisms. The normal and most convenient method of sterilizing domestic water supplies is the addition of one part in a million of chlorine gas. Water may be sterilized on a smaller scale by boiling, distillation, ultra-violet radiation, or the addition of chemical compounds which are toxic to living organisms.

The purity of the resulting water may be tested by noting the population numbers of *Bacillus coli,* a harmless bacterium found in the human intestine. If the water is in any way polluted this bacterium will be present.

The softening of hard water may be necessary in some areas if the water reaches the user by way of slightly soluble underground rock layers. This water will contain salts that prevent it from lathering well with soap. Such water also "furs up" pipes and boilers with a "scale" of carbonate or sulphate.

Hard water is of two types:

Temporary hard water, which is caused by dissolved salts of calcium and magnesium bicarbonate. These salts will be converted to insoluble carbonate by boiling. This water may therefore be softened by boiling. The addition of lime will also soften it.

Permanent hard water, which is caused by the presence of calcium and magnesium sulphates. They may *not* be removed by boiling. This type of hardness may, however, be removed by the addition of washing soda, sodium carbonate, or by the use of a modern "ion-exchange" water softener.

DISPOSAL OF WASTE

During community living man accumulates much that is useless and, in some cases, offensive and dangerous.

(i) Excreta, human and animal.

(ii) Domestic refuse, dry and wet.

(iii) Industrial wastes.

Together with the above, rain water and street debris must be removed. The methods employed for the disposal of sewage depend upon the presence of a convenient source of water.

(a) The *"water carriage" system* is a method employing sewage pipes, which carry all types of waste away from the home, factory and street to the sewage works or area of disposal. This is the modern town method, requiring an adequate water supply and a sewage works for rendering the material harmless.

(b) *Dry or conservancy systems* are employed where an inadequate or no water supply is available. These dry methods have many disadvantages which assist the spread of disease. Such methods are the use of the privy-midden, closets employing trenches, buckets and chemical bins, or the camp latrine.

The Water Carriage System of Sewage Disposal

The dry methods of waste disposal mentioned above are inefficient and unpleasant. The sewage has to be collected, transported and disposed of in some way.

The scheme of modern sewage disposal is as follows. A receptacle for the waste, such as a lavatory bowl or drain, is joined to the main street sewage pipe. The material is carried in water to a purifying plant where the following treatment takes place.

PRELIMINARY TREATMENT

The crude sewage is pumped into a tank which has a screen to trap papers, rags or any other large objects. The sewage then flows slowly along grit channels, where the grit settles to the bottom. The sewage next flows into a series of sedimentation tanks with sloping floors. After a few hours most solid material settles to the bottom as a *sludge*. The remaining liquid, the *effluent*, is then subjected to one of the following purifying methods:

The percolating filter method.

The activated sludge method.

The percolating filter method employs a large filter bed of coke, stones or clinker onto which the effluent is sprayed from rotating arms. The offensive organic material in the effluent is attacked by aerobic bacteria which form a film over the surfaces of the coke or gravel. The effluent trickles down through 2m of this filter. After the filter, the liquid is passed into a humus tank where all remaining solid matter settles out. The purified effluent is now passed into the river or sea.

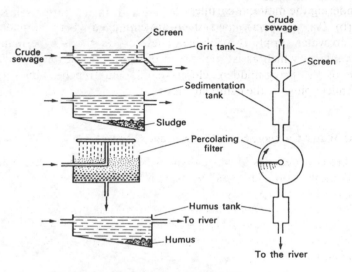

Fig. 99. Sewage disposal—percolating filter method.

The activated sludge method does not employ the large filter beds. The effluent from the sedimentation tanks is violently agitated by the forcing through it of compressed air, or by continuous stirring. The effluent is now passed to further settling tanks where a sludge forms on the bottom. A portion of this sludge, which is rich in bacteria, is pumped back into the crude sewage entering the first sedimentation tanks. The presence of these bacteria during the agitation process allows the organic material to be changed into less offensive matter.

The harmless effluent is discharged into the river. The sludge taken from the settling tanks is sometimes dried into "cakes" and sold as manure.

Where a dwelling is not connected to the main sewage pipes a *cesspool* is a common feature of the design. This cesspool, or septic tank, takes all the sewage and is periodically emptied by suction. The sewage is then transported in tankers to the sewage works. It is important that the design of the cesspool will prevent any offensive gases from reaching the house.

Domestic refuse is usually collected and sorted into useless and useful categories. It may then be tipped or incinerated.

CHAPTER 16

Man in the Community

MEN in towns no longer all survive on their own cultivated crops, livestock or livestock products, or the exchange of commodities. Modern man feeds from a limited supply of fresh food supplemented by preserved foods. Town man no longer exchanges or barters goods in order to obtain the things he needs. Modern man very often gives his time and muscular and mental energy in reward for a new commodity, money. Money is a token which man may exchange for the goods he needs. Modern man is either an employer or an employee, money being the reward of both occupations.

In the course of man's employment and community living he needs economic, medical and psychological protection. Socially conscious men construct institutions to cater for these needs. Community living involves:

1. Food hygiene and preservation.
2. Industrial health services.
3. National health services.

1. FOOD HYGIENE AND PRESERVATION

Food hygiene is important for obvious reasons. An infective person or animal contacting food will infect the food. Infective food will produce disease or discomfort in the consumer. The possible diseases transmitted by way of food and drink are discussed on pp. 230–4.

Persons handling food should always wash their hands first, especially after visiting the lavatory. Hands with cuts and sores should be well covered. Persons with any form of illness, such as sore throat, diarrhoea or cold, should not be permitted to contact food directly or indirectly by coughing and sneezing.

In order to prevent the contamination of food by micro-organisms, food should always be kept covered or preserved in the manner described as follows.

The utensils used to serve food, or from which food is eaten, should be clean, having been dried in a stream of hot air after rinsing. It is advisable that women who are dealing with food should wear their hair covered.

In order to demonstrate the presence of micro-organisms in the air or on the fingers, a small plate (Petri dish) of meat extract jelly need only be touched, and then left in an incubator for a few hours. Close examination of the plate later will reveal colonies of organisms growing across the nutrient jelly.

Food preservation is necessary if man is going to eat foods out of season which have not by that time been contaminated or decomposed.

Some methods of preservation are as follows:

 (i) Refrigeration.
 (ii) Canning.
 (iii) The use of chemicals.
 (iv) Curing.
 (v) Dehydration.

(i) *Refrigeration* methods depend on the fact that at low temperatures chemical reactions slow down and consequently micro-organisms are retarded in their growth. Refrigeration methods may damage foods if the freezing and thawing take place slowly. Rapid freezing is recommended. Micro-organisms are not necessarily destroyed by cold, but their growth and multiplication are inhibited.

(ii) *Canning* involves sealing food into an air-tight, partially evacuated container. It is then heated to very high temperatures to destroy the micro-organisms. Where the container is made of metal, precautions must be taken to prevent the metal from corroding.

(iii) *Chemicals* are sometimes used as preservatives, but their use is strictly controlled and limited. Some chemicals are: sulphur dioxide, nitrates and nitrites of potassium and sodium, and benzoic acid.

(iv) *Curing* meats is another method of preservation. Fish, ham and bacon may be cured by smoking with the fumes of burning oak, elm or beach shavings. This smoke impregnates the meat and destroys many micro-organisms that would hasten the decomposition of the meat. Saltpetre and salt may also be used to cure bacon quickly.

(v) *Dehydration* is the removal of water from the food to be preserved. The absence of water makes it difficult for micro-organisms to survive. These foods, usually finely shredded for ease of dehydration, should have water added to them before they are eaten.

2. INDUSTRIAL HEALTH SERVICES

The industrial health services had their origins in the Factory Acts of 1937, 1948, 1959 and earlier. These Acts increased the provisions for the health and safety of all types of employed persons. The first Factory Act was passed in 1833 when the government's power to inspect factories was enforced. The hours of work for children and women were limited by this Act. Today the industrial worker is protected against disease and accident: he is paid benefits in illness and helped to be rehabilitated after disease or injury.

The following are some of the provisions related to industrial health featured in the Acts:

Health and comfort. Adequate ventilation, lighting and temperature of working rooms. Prevention of overcrowding. Seats available for women workers, rest rooms and suitable breaks in which to rest. Good washing facilities and lavatories. Painting of internal walls at regular recorded intervals. Disposal of any rubbish accumulated in the work. Provision of canteens.

Safety. Notification of the Chief Factory Inspector of any illness resulting from work with lead, mercury or phosphorus compounds. Provision of first-aid facilities, industrial nurse and medical officer. Provision of protective clothing, machine guards, adequate fire escape facilities. The prevention of noxious fumes and dusts from polluting the air. The control of the employment of young persons and the number of hours they may work.

Compensation for disablement. Injuries sustained during employment are eligible for financial compensation and treatment under the National Health Service. Re-employment for the disabled was the concern of the Disabled Persons (Employment) Act, 1944.

Some occupational diseases associated with particular trades are listed as follows:

Industrial dermatitis occurs commonly among chemical workers, engineers and technicians using X-ray or radioactive sources. The skin becomes inflamed and sore through exposure to noxious substances.

Silicosis results from prolonged inhalation of fine particles. This is one of the more common *pneumoconioses* to which miners, glass makers and potters are subject. The patient becomes short of breath and coughs frequently. The lungs contain nodules formed around the retained particles.

Miner's nystagmus is a side to side oscillation of the eyeballs found in mineworkers with long employment in dim underground light.

Raynaud's disease, known as "dead fingers" or "white fingers", may be produced in workers who have for many years handled rapidly vibrating tools. The fingers become bloodless and insensitive.

Blindness may occur if a worker is subject to high temperatures for long periods. This produces cataract blindness.

Deafness may be caused by excessive noise.

"The bends" may result if a worker is employed under high pressure, such as in deep sea diving or in a caisson for underwater construction. At these pressures nitrogen dissolves in the blood; if the pressure is reduced too quickly the nitrogen bubbles appear in the blood and cause violent pain. This condition is treated by slow decompression in a decompression chamber.

3. THE NATIONAL HEALTH SERVICE

An Act to provide a comprehensive health service for England and Wales became law in 1946. The services under this Act began in 1948. From this time the Minister responsible for Health became responsible to the government for an organization directed at curing and preventing disease. Previous to this Act the services of the medical profession were given voluntarily or for a fee.

Under this Act hospitals, consultants, specialists, nurses, midwives and general practitioners were united under the National Health Service. Many local services were also initiated or improved, such as health visitors, ambulance services, home helps and home nursing.

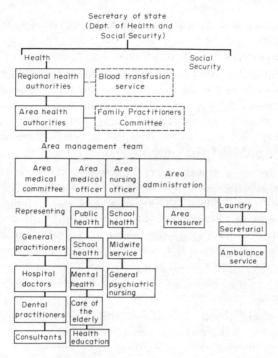

The administration of the National Health Service is organized as shown in the diagram above.

It is the function of the local health authority to appoint a Medical Officer of Health, who is the chief administrator of the Public Health Department.

The Medical Officer of Health supervises the following:

Health Visitors—usually state registered nurses who attend to the needs of mothers, children, old and bed-ridden people, the sick and the mentally ill or deficient. Their duty is to administer, report and advise on all health matters.

Maternity services—antenatal clinics, midwifery services, child clinics.

Home nursing or District Nurses who nurse the sick at home.

Immunization clinics which give immunizations to prevent diphtheria, poliomyelitis and whooping cough.

School health services which are provided by the Local Education Authority. School clinics and nurses provide a service for all school children in each area.

The Medical Officer of Health is responsible for all aspects of health and hygiene within his area. The above are but a few of the duties supervised by the Medical Officer of Health.

ATOMIC RADIATION AND MAN

Modern industry makes use of radioactive materials and modern warfare involves the testing and possible use of thermonuclear devices. Both aspects may result in the exposure of people to higher radiation than is normal.

Radioactivity refers to materials which emit rays (atomic radiations) capable of penetrating not only human flesh, but also many inanimate objects such as wood, concrete and metal.

The effects of these radiations are many, but four possible effects are listed as follows.

Cancers of the blood (leukaemia), the skin or the internal organs may result after prolonged exposure to radiations such as X-rays.

Radiation sickness is a general name given to a collection of symptoms following exposure to radiation above a certain value.

The radiation may de-nature enzymes, antibodies, vitamins and many other vital chemical units. Vomiting, indigestion and a tendency to skin and intestine infections are predominant after severe doses of radiation. A white blood cell count is taken after exposure to radiation as the numbers of these cells increase after such radiation and thus give some indication of severity of the dose incurred.

Abnormal child development may result if a pregnant woman is subjected to a large dose of radiation. The embryo is more subject to these radiations during the early stages of development. Deformities of body and mind may result from irradiation during intra-uterine life.

Abnormal future generations may arise if gene structures are altered (mutated) by a bombardment with atomic particles. The mutations that might appear in future races can only be guessed at, but mutations are usually "abnormal".

The main sources of atomic radiations in our lives are:
(a) *Cosmic radiation* from outer space and from rocks.
(b) *X-ray machines.*
(c) *Radium* used in cancer treatment.
(d) *Atomic power stations* and allied research institutes.
(e) *Luminous paint* on watch faces.
(f) *Testing of nuclear weapons.*

Strict measures are adopted by all establishments employing radioactive materials, in order to prevent personnel from receiving harmful doses of radiation.

CHAPTER 17

Disease

THE diseases to which human beings are subject may be grouped as follows:
1. Hereditary diseases.
2. Birth injuries or infections.
3. Infectious diseases (communicable).
4. Organ malfunctioning; age and injury.
5. Deficiency diseases.
6. Tumours.
7. Occupational diseases.
8. Mental illness and subnormality.

1. HEREDITARY DISEASES

Hereditary diseases are transmitted from parent to offspring as a result of a particular genetic pattern or "blueprint". Such disorders are very rare because they are genetically "recessive". One example is *haemophilia,* which is sex-linked with "maleness". Other examples are the *muscular* dystrophies.

Mongolism is a condition of mental and physical retardation resulting from a change in the chromosome number. This condition is "genetic", but it is not hereditary. Mongols are usually sterile.

2. BIRTH INJURIES OR INFECTIONS

Birth injuries to the head may produce a mentally defective child.

226

Birth infections are not hereditary diseases but are diseases transmitted from the mother to the child during pregnancy. Such a congenital disease is *congenital syphilis.*

3. INFECTIOUS DISEASES

The foregoing disorders are determined before birth. The infectious diseases to be mentioned below are communicated from person to person in one way or another. These diseases may be:

Pandemic—affecting a large percentage of the population of many countries at the same time.

Epidemic—affecting a large percentage of a community at the same time.

Endemic—a disease characteristic of a certain group at a certain time.

A study of communicable disease must involve a knowledge of:

(i) The natural or acquired immunity of persons.

(ii) The causal agent of the disease.

(iii) The method of entry to the body.

(iv) The mode of transmission from person to person.

(v) The methods of control and prevention.

(i) Natural or Acquired Immunity to Disease

Immunity is the ability of a person's tissues to resist a particular infectious disease. This immunity may be present at birth, that is, *natural.* The ability to resist infection may be *acquired* as a result of a previous attack of a certain disease, or of repeated vaccinations with mild disease-producing substances.

NATURAL IMMUNITY

Organisms which cause disease in man are known as *pathogenic.* There are persons who are not affected by certain pathogenic organisms. They are said to be *immune* to that organism. This immunity is a genetic disposition of that particular individual.

ACQUIRED IMMUNITY

Acquired immunity may be:

(a) Acquired naturally.

(b) Acquired artificially.

(a) Immunity to certain disease-producing organisms may be acquired after being exposed to small doses of a pathogenic organism. Again, on recovery from certain infectious diseases (measles, chickenpox, smallpox) it is unusual for that person to suffer a second attack. The individual has a *naturally acquired* immunity.

(b) Immunity of an artificially acquired nature may be:

Active immunity, which means that after inoculation with a certain organism the individual produces his own protective substances.

Passive immunity refers to the immunity resulting from an injection of anti-toxin. The individual does not produce his own protective substances.

Immunities of the above two types are used as protection against certain diseases.

Active immunization is used as protection against smallpox, poliomyelitis, tuberculosis, tetanus, typhus, enteric fever, yellow fever and whooping cough.

Passive immunization may be used to combat diphtheria, measles and tetanus. This type of immunization is used when a patient is actually suffering from a disease, because active immunity takes time to establish itself.

(ii) The Causal Agent of Disease

Communicable diseases may be caused by any of the following unicellular organisms.

(a) *Cocci* are spherical bacteria which may associate in chains, bunches or in pairs.

> Chains: streptococci, e.g. scarlet fever.
>
> Bunches: staphylococci, e.g. boils, acne.
>
> In pairs: diplococci, e.g. pneumonia.

Bacilli are rod-shaped bacteria, some of which have the power of movement. Some of these bacilli will form themselves into protective

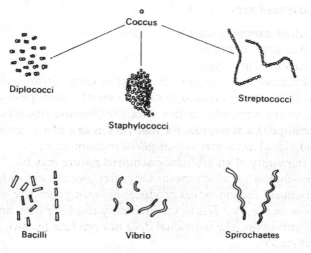

Fig. 100. Bacteria.

spores during hard times, e.g. tetanus bacilli, typhoid bacilli or tubercle bacilli.

Spirilla (the Vibrios) are comma-shaped bacteria, e.g. cholera.

Spirochaetes are spiral-shaped bacteria which are usually very motile, e.g. syphilis.

Pathogenic bacteria are parasites which live on and receive their nourishment from another living organism (their host). Pathogenic bacteria produce substances that are poisonous to the living tissues. These *toxins* stimulate the host to produce *anti-toxins*.

(b) *Viruses* are organisms too small to be observed with a light microscope. They are studied beneath the electron microscope. Such organisms reproduce within the living cell.

Some examples of virus-caused disease are measles, chickenpox, smallpox and influenza.

(c) *The protozoa* are a group of unicellular animals which, if disease-causing, are parasitic within the human body, e.g. amoebic dysentery, malaria and sleeping sickness.

(d) *Fungi* are plants which are unable to manufacture food by photosynthesis because they possess no chlorophyll. Disease-producing fungi live on the human tissues, e.g. athlete's foot and ringworm.

(iii) Method of Entry to the Body

Pathogenic organisms may enter the body by way of:
(a) The respiratory system.
(b) The alimentary canal.
(c) The skin.

(iv) Mode of Transmission from Person to Person

The causal agent of disease is carried to the point of entry to the body by way of:
(a) *The air,* which carries organisms from one host to another.
(b) *Food and drink* taken in, which may carry pathogenic organisms.
(c) *The skin,* which when punctured by infected animals or objects allows the entrance of pathogens.

(a) BY AIR

Pathogenic organisms may be transmitted from person to person in droplets of saliva or sputum expelled from the mouth or nose in coughing, sneezing and talking. A sneeze will propel infectious droplets across a room.

Diseases transmitted in this way are: the common cold, influenza and measles (virus diseases); pneumonia, diphtheria, whooping cough, tuberculosis and scarlet fever (bacterial diseases).

(b) BY FOOD AND DRINK

The pathogenic organism may gain entry to the alimentary canal after infected food and drink have been taken. The food may be contaminated in one or many of the following ways:
Finger–food contact.
Droplet–food contact.
Sewage–domestic water contact.
Infective persons in contact with food and water (carriers).
Infected animals in contact with food and drink.
Finger–food contact. Food may become infected as a result of infec-

tive fingers handling it. Infective fingers placed into the mouth or onto food will serve as one mode of entry to the alimentary canal for pathogenic organisms.

Clean hands only should handle food. A septic wound should always be cleanly dressed because the organisms causing sepsis (staphylococci) may cause violent food-poisoning.

Droplet–food contact. Droplets of vapour contaminated with pathogens may infect food if it is exposed to the atmosphere.

Sewage in contact with the domestic water supply. This contact may permit the transfer of living organisms from the infective excreta to the water used for drinking or washing. Any water used for drinking and washing should be purified in order to kill pathogens capable of causing cholera and other enteric fevers.

Infective persons in contact with food and water. Such persons spread pathogens without being aware that they are doing so. These people are often referred to as "carriers". A person incubating or recovering from a disease may harbour bacteria capable of infecting the water or food handled by that individual. Dysentery, typhoid and food poisoning may be quickly transmitted to others by way of food and drink which has been in contact with a human carrier.

Insect carriers in contact with food and drink. Insects are capable of spreading the causal agents of such epidemic diseases as typhoid, dysentery and cholera. Many minor but unpleasant enteric disorders may also result from eating food upon which flies have been walking and feeding. The hairy legs of these animals provide an excellent vehicle for bacteria picked up from refuse and dung heaps.

The common house-fly and its fellow, the bluebottle, should be exterminated by all possible means. To reduce the risk of a fly's carrying disease-producing organisms, all refuse and excreta should be covered or, preferably, burnt.

Infected animals may be directly or indirectly responsible for transmitting disease to human beings. The flesh of animals may harbour some stage of the life cycle of a disease-causing parasitic worm or fluke. Meat must be inspected before sale to reveal the intermediate stages of the following parasites:

Taenia solium, the pork tapeworm, has an intermediary in its life cycle which lodges in the muscle of a pig, causing "measly pork".

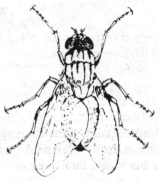

The eggs

The larva

The pupa

Adult fly

FIG. 101. The house-fly.

When this meat is eaten a worm up to 4m long develops in man's intestine and feeds on his food.

Taenia saginata, the beef tapeworm, spends a stage of its life cycle lodged in beef muscle. This organism will become a parasitic worm in the gut of a human being.

Trichinella spiralis is a smallish worm which dwells in man's muscles, causing discomfort and sometimes death. This parasite may be ingested as a result of eating contaminated pork. These meats must always be properly cooked before eating. Suspect meat should be burnt.

Cow's milk may be contaminated with bacteria causing *tuberculosis.* Dairy herds should be tuberculin-tested. Milk is always pasteurized to reduce the risk of infection.

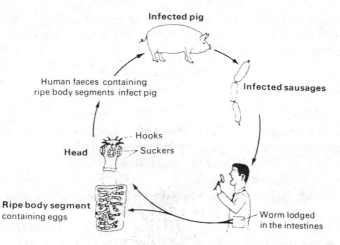

FIG. 102. The pork tapeworm.

Sausage meat and some canned foods may harbour organisms which produce poisons in badly preserved material. These poisons cause an uncommon but often fatal type of food poisoning called *botulism*.

Food poisoning usually refers to those gastro-intestinal disorders resulting from the swallowing of food or drink contaminated by bacteria or their poisons. The mucosa of the stomach and intestines becomes irritated.

Salmonella species of bacteria contaminate food by contact with animals and birds or by contact with carriers of the bacteria. Meats, cream, eggs (especially duck eggs) and egg meals are common carriers of this group of bacteria, even though the food looks normal.

Salmonella typhic (typhoid bacillus) is a bacterium which parasitizes man, being taken in by contaminated water or food. Flies can spread this bacillus. Typhoid is caused by the related bacteria of food poisoning but it is an extremely dangerous infectious disease rather than a "food poisoning".

Staphylococci strains of bacteria grow in food and produce poisons. Cooking the food can kill the bacteria but the poisons may remain behind, causing intestinal irritation. The food can be contaminated by a septic wound or sore on the finger or from the nose or skin of a carrier. A rare but fatal form of this type of food poisoning has been mentioned earlier—botulism.

(c) BY WAY OF THE SKIN

Disease-causing organisms rarely penetrate the skin alone; they are usually introduced into the body by one of the following methods:
Wounds or the hair follicle.
Animal bites.
There are certain worms, however, which have young stages capable of boring an entry through the human skin.

Blood flukes (*Schistosoma*) have young stages that bore through the skin of persons wading in infected water. The worm lives in the blood and produces a disease known as *bilharziasis*.

Hook worms also have a young stage that is capable of penetrating man's skin.

Wound infection

Septic conditions may be produced if the skin is broken and staphylococci multiply in the nutritive tissues of the body. The same conditions may arise as a result of bacteria entering the deeper tissues by way of hair follicles. Wound dressings should always be sterilized. *Tetanus* may originate through an open wound. This is caused by the tetanus bacillus (*Clostridium tetani*) which may be in the spore condition in soil.

Infective animal bites

An animal may puncture the skin and force into the tissues pathogenic organisms resting on the skin or present on the mouth parts of the animal.

Dog bites may infect a person with the virus causing rabies. This fatal virus will be present in the saliva of infected dogs and cats.

Rat bites may produce an unpleasant fever, caused by a virus present in the saliva of the infected rat.

Mosquito "bites" are in fact injections of the insect proboscis into the skin. Only the female sucks blood for her meal, and in doing so she vomits out saliva which acts as an anti-coagulant. The saliva may contain organisms causing certain diseases. There are three important

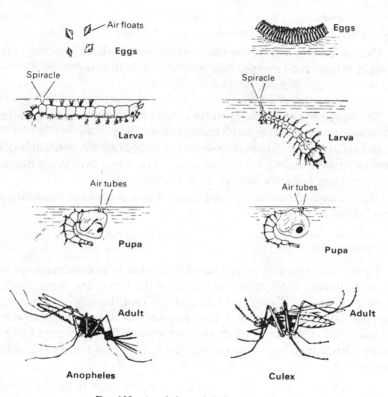

FIG. 103. *Anopheles* and *Culex* mosquitoes.

mosquitoes to consider here, only two of which are carriers of disease-producing organisms. The mosquito common to Britain is a species of *Culex*. This mosquito does not carry pathogenic organisms harmful to man. It may be distinguished from the other two mosquito types as shown in Fig. 103.

Mosquitoes of the species *Anopheles* are the sole transmitters of human *malaria*. The female mosquito feeds on the blood of a person and in so doing transmits the protozoan *Plasmodium* from person to person. This species of mosquito actually carrying the plasmodium is found only in warmer climates.

The mosquito *Aedes aegypti* transmits a virus in its saliva which produces *yellow fever*. This severe jaundice type of disorder occurs only in tropical forested areas of America and Central Africa.

The dengue is a virus disease transmitted by mosquitoes in warmer countries.

Elephantiasis is a disease produced by a microscopic worm which causes obstructions in the lymphatic vessels. The lower limbs of the infected person swell up with tissue fluid unable to return to the blood circulation. These *filarial worms* are transmitted by mosquitoes in warmer Eastern countries.

The life cycle of the mosquito	Methods of control
1. Eggs laid in water. 2. Larvae live in water, breathing by means of spiracle at water surface.	Clear away all swamps and receptacles capable of holding water. Drain all pools, ditches, etc. Cover water surfaces with oil. Introduce larvae-eating fish into pools, etc.
3. Adults are limited in flight range and are sensitive to DDT and other insecticides.	Clear vegetation and possible breeding grounds in wide area around human habitation. Use mosquito nets and insecticides. Remove infected malarial patients from possible contact with feeding mosquitoes.

Rat flea bites may spread a condition known as the *plague*. When rats become infected by the plague bacillus, the fleas which live upon the blood of the rat also become infected. When the rat dies, the flea abandons the rat and, if in close contact, takes to the blood of man. The plague may easily spread to epidemic or pandemic proportions (such as the Black Death in the fourteenth century or the Great Plague of London in 1665), and may only be controlled by rat extermination and vaccination.

Body louse bites may transmit a condition known as *typhus*. The louse, infected with a microscopic bacteria-like organism, *Rickettsia,* ruptures the skin with its mouth parts and so allows the organism entry to the body tissues. Scratching the skin, an irritation may assist the entry of *Rickettsia.* The faeces of the louse may remain infective for some time and are therefore a possible source of infection. The destruction of lice is the most certain method of controlling typhus. DDT is effective against this insect.

(v) Methods of Control and Prevention

If a disease of an infectious nature is seen to occur in a member of the population, it is important that this fact be reported to the necessary authorities for a number of reasons:

(a) To discover the origins of the illness.

(b) To trace and isolate carriers.

(c) To take the appropriate steps to prevent spread, by vaccination and destruction of infective material.

Diseases which must by statute be reported to the local health authority are called *notifiable diseases.* The main diseases concerned are listed as follows:

Smallpox	Malaria	Diphtheria
Typhoid	Tuberculosis	Measles
Typhus	Dysentery	Pneumonia
Poliomyelitis	Cholera	Food poisoning
Plague	Scarlet fever	Leprosy
Anthrax	Whooping cough	Encephalitis

In order to prevent the spread of pathogens throughout the population, objects that are likely to be infective are treated in such a way that the offending organisms are destroyed. When an object is free from pathogens it is said to be *sterile.* The article is then *aseptic.*

A disinfectant is an agent that kills pathogenic organisms.

An antiseptic is an agent which tends to prevent the growth of organisms causing sepsis in wounds.

Sterilization methods include the following:

Dry heat

Burning infected objects is the most satisfactory method of destroying pathogens.

Hot air sterilizers at approximately 160°C (320°F) will kill organisms on objects that will not be damaged at this temperature.

Damp heat
This is more satisfactory as a sterilization method because water vapour will more easily penetrate the organism and coagulate its protein composition.

Boiling water will sterilize satisfactorily if the infected object is immersed for half an hour.

Steam sterilization is also a very satisfactory method of killing bacteria and spores.

Chemical agents
Mercury salts such as mercuric chloride are excellent germicides, but are poisonous to man.

Oxidizing agents such as hydrogen peroxide yield oxygen, which promotes the destruction of pathogenic bacteria.

Iodine and chlorine compounds are used to sterilize the skin and water supplies.

Phenol or carbolic acid as a 1:100 solution is a powerful germicide.

Spirits such as surgical spirit (alcohols) are useful sterilizing agents.

Radiation by gamma rays is an effective way of sterilizing surgical instruments.

4. ORGAN MALFUNCTIONING; AGE AND INJURY

Under this heading may be included disorders resulting from malfunctioning of the endocrine glands.

Diabetes is the inability of the pancreas to produce insulin.

Goitre is the abnormal functioning of the thyroid gland.

Myxoedema results from an underactive thyroid gland in adult life.

Cretinism results from an underactive thyroid gland in infancy and produces a retarded individual.

Sexual disturbances, in physiology and behaviour, may be produced by abnormal conditions existing in the adrenal glands or sex glands or in the general hormonal balance.

Prostate enlargement in older men necessitates surgery as this gland exerts a pressure on the bladder and urethra.

Disorders of the uterus may follow the menopause, necessitating surgery of one sort or another.

Older persons tend to develop symptoms of disorder produced by a combination of many upsets. The general senility of some may be the result of harsh conditions throughout their life. Failing eyesight, hearing and mental acuity are the results of wear and tear. People age differently and at different rates; this is most probably genetically determined.

The influences of loneliness, of a general inability to cope as one used to and of some pain tend to age or speed up degenerative changes.

Injury may result in temporary or permanent disorders such as deafness or blindness.

The swallowing of corrosive substances, burning the skin, breaking bones and wounding also contribute to the ill health of the body.

5. DEFICIENCY DISEASES

Reference has already been made to disorders resulting from badly planned diets. Diseases such as those following may be produced as a result of vitamin deficiencies.

Xerophthalmia is a cornification of the cornea seen in young children deficient in vitamin A. The welfare of the skin also relies upon this vitamin.

Beri-beri and pellagra are diseases resulting from diet deficiencies of vitamin B.

Scurvy results if vitamin C is consistently absent from the diet.

Rickets and other bone deformities result if vitamin D is excluded from the diet.

6. TUMOURS

The occurrence of abnormal growths in various organs of the body is categorized under the group name of *cancer*. Cancers are various, and their causes are hidden among a multitude of interacting factors, such as environment, chemical irritants, glandular unbalance, psychological disposition, etc.

Cancers are produced as a result of abnormal changes in the cell nucleus which cause the cells to deviate from their normal course of activity.

Chemical agents capable of causing *carcinomas* (cancers) are known as *carcinogens*. Many carcinogenic chemicals have been isolated from common substances with which we have daily contact. Cigarette tobacco tars have been shown to contain carcinogenic chemicals, one example being 3, 4-benzopyrene.

7. OCCUPATIONAL DISEASES

An individual's occupation may bring him into an environment which is likely to cause disease after a period of time.

Particles of sand, flour, chemicals, coal, china clay or glass may cause lung inflammation of varying severity. These are the *pneumoconioses*.

These particles may be accumulative poisons (which in time become effective if absorbed in sufficient quantities), or carriers of infective organisms. Another disorder which may arise during working hours is some form of *dermatitis*. Contact with chemical agents, e.g. hair-perming lotions, may irritate some skins to the extent of producing disease.

Employers generally recognize the existence of these disorders and adopt the necessary precautions, such as insisting on the use of dust masks and other protective clothing.

Stress diseases are also induced by the working environment. These may result from psychological dispositions, combined with factors such as responsibility and the frustrations of a competitive existence. Peptic ulcers and coronary conditions seem most frequently to affect those persons in middle years with posts of responsibility.

8. MENTAL ILLNESS AND SUBNORMALITY

A mental subnormality refers to those conditions defined by the Mental Health Act of 1959 as a state of incomplete or arrested development of the mind. Persons of subnormal type need special care and training. This defect usually arises at birth or before.

A mental illness is extremely difficult to define because if we describe it as "an abnormal working of the mind" we must first know what is the "normal working of the mind". This we cannot easily do. A *psychologist* is a student of the "normal" workings of the mind. A *psychiatrist* is a medically trained person studying the "abnormal" functioning of the mind.

Apart from the problem of definitions of normal and abnormal is the even greater problem of what is "mind". Mental illness may be roughly grouped into the following three categories:

The neuroses are often emotional disorders brought on by some *real* factor in the life of the person. The person becomes anxious or depressed and so his behaviour changes. The sufferer is aware of his problem but may take avoidance tactics rather than face the real problem. He may relieve his tension by alcohol or drugs. Physical illness may be brought on by prolonged neurotic suffering or suicide may be a last resort to overcome the problem.

The psychoses are those categories of suffering where the emotional disruption is brought on by some *unreal* factor in the life of the person. The psychotic may be unaware of anything wrong as he may live in a fantasy world. He differs from the neurotic sufferer in being detached from reality and is thereby more of a problem for treatment. A common psychosis is *schizophrenia* (split mind) where the person alternates between normal and abnormal behaviour.

The psychopathic personality is another category of behaviour disorder which is not easy to define. In most respects these people seem normal except that they behave in an anti-social manner, as criminals, drug addicts or tramps. This is not so much an illness for the person concerned but an unusual way of life.

From this brief summary it can be seen how much of a problem it is to define the terms used in psychology. Naming is not explaining and so care must be exercised in using such terms.

Mental Health and Mental Hygiene

Mental health and mental hygiene are comparatively recent terms in our daily vocabulary. They represent not new problems, but a new and enlightened attitude to our inner experiences, emotions and behaviour. The days of being "possessed by the devil" or being "born evil or in sin" are fortunately passing. Despite the increase of our knowledge concerning the "inner life", a social stigma is still strongly attached to "being mentally ill". The mentally ill in all probability outnumber the physically ill, but they receive something like social rejection, which is the worst treatment and in many cases the one *cause* of their illness. Society shows its mental ill health as delinquency, criminality, perverted behaviour, alcoholism, the neuroses and the psychoses. All are expressions, in different degrees of mental ill health, and consequently require similar sympathetic consideration. The causes of much disturbed behaviour, where not obviously biochemical, are thought to be buried in the past. The essential *mothering* of the human child should be by the *parents,* in the *home.* Attempts to provide substitutes for either are thought to be the causes of much ill-adapted behaviour. Mental ill health must be prevented, and this is the field of mental hygiene. In the briefest terms possible, mental hygiene may be reduced to "being good parents" and an understanding and sympathetic adult.

CHAPTER 18

Some Practical Work

THE practical exercises suggested here will vary from very simple to quite advanced. Some will demand little or no apparatus, others will require more complex apparatus but no more complex than that found in the average teaching laboratory of school or college.

The student is advised to attempt as many of these exercises as possible, particularly those students intending to follow on later to advanced biological scienses, the "A" level or Technical Certificates in Biology.

EXERCISE 1. EXAMINING THE HUMAN BODY

REQUIREMENTS

A mirror and hand-lens, a ruler.

PROCEDURE

(i) *Examining the head*

Look into a mirror and notice the following:

(a) *Two eyes* which are placed so as to allow stereoscopic vision. Move nearer the mirror whilst focusing attention upon the nose. Stop at a point where focusing becomes difficult. Measure this distance— does it differ from person to person? Notice the two eyelids with eye-lashes in regular movement to clean and protect the front of the eye. Look at the eyebrows, ridges of hair which may be so placed to stop sweat running down the forehead and into the eyes. Eyebrows differ from person to person.

(b) *The nose* has a pair of nostrils, their openings being guarded by longish hairs.

Place the nose near the mirror, breathe on it and notice what happens.

Rub the bridge of the nose with the mirror surface and notice the greasy streak produced. The nose has many oil glands which make the nose shiny. This grease may trap dust, preventing it going further into the eyes.

Look at other people's noses and notice how different they all are. These nose characteristics we inherit from our parents. Look at your parents' noses.

(c) *The mouth* is bounded by very mobile lips which are sensitive and are used in eating and speech. Why should the lips be such mobile and sensitive structures?

Open the mouth and examine the teeth; count those on the top jaw and on the bottom jaw. Stick the tongue out and draw it back in again, rubbing the surface along the front teeth. The surface feels "bumpy" because of taste organs.

Can the tongue be curled up into a tube? Check with other people to see if they are able to do the same. The ability to do this is inherited. Check to see if your parents or any of the family are able to do this.

Open the mouth wide, lower the back of the tongue and notice the fleshy structure hanging down. This structure will be considered in the chapter on the alimentary canal.

(d) *The cheeks* are extremely mobile areas of flesh which are frequently used in social communication. These areas of skin are moved in certain ways to express our feelings—smiling, fear, hate and so forth. We do not need to be taught how to use these face muscles, as can be seen with persons blind from birth. Try to imitate different emotions using your face muscles. Notice how some people have more expressive faces than others.

(ii) *Examining the hands and arms*

(a) *The hands* are very flexible structures, being specially constructed in man to enable him to hold objects and to use tools. The thumbs can be placed in front of each finger, making the hand an extremely accurate and useful limb.

(b) *The forearms* are also interesting in man because they can be rotated.

Stretch out the arm with the palm of the hand facing upwards. Hold the elbow firmly. Rotate the forearm and bring the palm of the hand to face downwards. Very few animals can do this with their forearms! Look at the hairs on the arm with a hand-lens. How many are there on a square inch of skin? From this brief examination of the human body the following important human characteristics have been noted:

  (i) Stereoscopic vision.

 (ii) The ability to communicate verbally and facially.

(iii) The specialized forelimb.

(iv) Hair.

EXERCISE 2. USING THE MICROSCOPE

REQUIREMENTS

A microscope, a microscope lamp (unless the light source is built in), a microscope slide, some soft cleaning tissues.

THEORY

For the microscope to be used with the best results the glass lenses and the light source need attention before use.

PROCEDURE

Carry out the instructions given below and microscope work will become easier with less time wasted on searching for the specimen.

(a) Sit at a bench with the microscope at a convenient height to allow the eye to be put on the eyepiece without difficulty. Do not bend the microscope forward, if it has a joint to allow this. Fluids will run off the microscope.

(b) Feel for the double-sided mirror with the left hand and adjust it so that light from the lamp is projected onto it and reflected up into the microscope. Use the flat side of the mirror.

(c) Using the illustration below identify the following parts of the microscope: the condenser lens and its focus screw, the iris diaphragm and its control lever, the stage, the objective lenses, the fine and coarse adjustment screws, the body tube, the eyepiece.

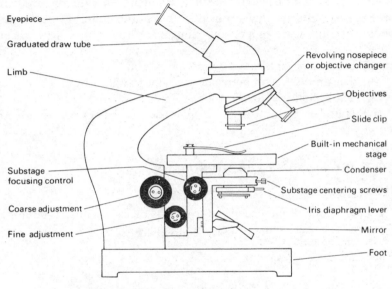

Eyepiece

Graduated draw tube

Limb

Revolving nosepiece or objective changer

Objectives

Slide clip

Built-in mechanical stage

Substage focusing control

Condenser

Substage centering screws

Coarse adjustment

Iris diaphragm lever

Fine adjustment

Mirror

Foot

FIG. 104. The student microscope.

(d) With a soft paper tissue gently wipe the lenses clear of dust. Try to avoid lifting out the eyepiece. Dust particles in a microscope can look as big as oranges!

(e) Turn the coarse adjustment screw until the objective lenses move. Twist the set of lenses until the lowest magnification object lens is in the action position. Turn the coarse focus further until the objective nearly touches the stage.

(f) Put a very small piece of dampened paper tissue onto a glass slide. Slip the slide beneath the low-power objective lens on the stage. Now we are ready to examine the magnified tissued fibres.

(g) Look down the eyepiece: you see light only. Turn the coarse focus slowly to raise the body tube up towards the eye. The fibres of the paper come into focus. If the light is too bright reduce it by turning the condenser lens screw to move this lens downwards. Adjust the angle of the mirror until the illumination gives a clear image.

(h) A point to remember:

NEVER LOOK DOWN AS YOU TURN DOWN.

Slides can be "drilled", glass can break and lenses damaged if the focusing is done by moving towards the specimen. Get into the habit of focusing up.

(i) To magnify to a greater degree twist the objective lenses around until the next highest lens comes into operation.

There is one lens with × 90 (or more) engraved on it; this is used immersed in oil: it is an oil immersion lens (considered later).

(j) A rough estimate of the magnification of a microscope can be worked out by multiplying the magnification of the eyepiece by the magnification of the objective lens.

EXERCISE 3. EXAMINING A BODY CELL

REQUIREMENTS

Microscope, microscope lamp (or source of light), microscope slide, cover slip, mounted needle, beaker, water, rubber teat pipette.

THEORY

Most parts of our bodies are made up of cells. These cells are different in their structure from one part of the body to another. Their structure depends upon their function. A cell taken from the heel of the foot would be expected to be different from one taken from inside the mouth.

PROCEDURE

(a) Set up the microscope ready for use with the low-power lens.
(b) Clean a microscope slide so that it is free from dust.

(c) Wash one finger so that it is clean enough to be put inside the mouth. Scrape the finger around the inside of the cheek.

(d) Smear the skin scrapings from the inside of the mouth onto the cleaned slide. Keep the smear moist by adding a small amount of water.

(e) Put this sample under the microscope and focus up. Search the smear for cells. These flat scale-like cells may be difficult to find; adjust the mirror and lighting from time to time.

(f) If the smear begins to dry up then the cells become wrinkled and useless for examination. For this reason we normally cover our specimens with a glass cover slip (next exercise).

EXERCISE 4. STAINING A CELL NUCLEUS

REQUIREMENTS

A microscope, a microscope lamp, a microscope slide, a cover slip, a seeker, methylene blue stain, a piece of paper tissue.

THEORY

A blue stain called methylene blue will stain cell nuclei blue. A stained nucleus will stand out more clearly against the background of cytoplasm.

PROCEDURE

(a) Make a buccal smear (mouth cell smear) as done previously. Keep the cells moist with water.

(b) Lower a cover slip onto the smear as shown in the diagram on p. 249.

(c) Put a drop of methylene blue against the edge of the cover slip. Allow the stain to run under the cover slip. This may be speeded up by touching the edge of the slip on the side opposite the stain with a piece of paper tissue. This should pull the stain through.

(d) Examine the smear for stained cells under the microscope.

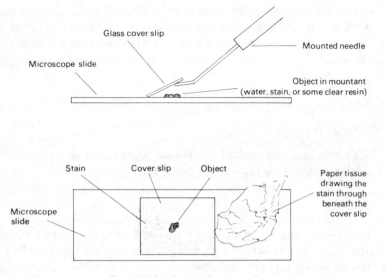

FIG. 105. Laying on the cover slip.

EXERCISE 5. EXAMINING STRIPED (VOLUNTARY) MUSCLE FIBRES

REQUIREMENTS

A microscope and lamp. A slide and cover slip. Two mounted needles. A small piece of muscle (meat).

THEORY

Voluntary muscle taken from the skeleton is distinguished from involuntary muscle by having striped fibres.

PROCEDURE

(a) Put a small piece of muscle (not much bigger than a few sugar grains) onto a slide. Keep it in water.

(b) With a pair of needles shred the muscle up a little in order to separate the individual fibres.

(c) Lower a cover slip onto the muscle fibres.

(d) Examine the slide for muscle fibres. It may be necessary to search for a good specimen. Adjust the light to show up the stripes on the muscle.

EXERCISE 6. EXAMINING THE SKELETON

REQUIREMENTS

A fully articulated or half-disarticulated skeleton, real or plastic. A labelled diagram of the human skeleton.

THEORY

The skeleton acts as a protective framework covering the vital organs of the body. It has attachment points for muscles. Blood cells are produced within the bone marrow.

PROCEDURE

One of the best ways to learn about the skeleton is to handle the individual bones and to attempt drawings of them.

(a) Commence by looking at the joints on the skeleton and classifying them as to type.

(b) Examine the long bones of the arm and leg. Look for muscle attachment points.

(c) Examine the flat bones of the pectoral and pelvic girdles. Locate the sockets for the arm and the leg.

(d) Examine the bones of the vertebral column and relate their structure to their functions (for example, lumbar vertebrae take the great back muscles).

(e) Examine the rib-cage and its attachment points with the vertebral column.

(f) Look for cartilage areas of the skeleton and suggest a reason why cartilage and not bone.

(g) Examine the skull closely, taking particular notice of the openings which penetrate the bone to carry nerves or blood-vessels.

Make a note of the names of the more obvious bones of the skull. Look for sutures.

EXERCISE 7. TESTING FOR CARBOHYDRATES

REQUIREMENTS

Bunsen burner, spatula, rack of pyrex test-tubes, watch glass, beaker 250 cm³, tripod and gauze. Carbohydrate samples, e.g. starch, glucose and sucrose. Benedict's reagent, Fehling's reagents. Dilute hydrochloric acid.

THEORY

Starch is a polysaccharide. It is insoluble in water and forms a gelatinous mass in warmed water. It can be digested by salivary amylase (ptyalin) to produce glucose sugar.

Glucose is a monosaccharide, soluble in water.

Sucrose is a disaccharide, soluble in water. It is made up of two monosaccharides linked together, glucose with fructose. It can be digested by an enzyme, invertase (or sucrase).

All these substances leave a lot of carbon on a spatula when heated strongly over a flame.

EXPERIMENTS

Starch test. Put a spatulaful of potato starch into a test-tube. Half-fill the tube with water. Bring the water to near boiling point in order to rupture the starch granules. Add a few drops of iodine solution. Shake the tube and notice the blue–black–purple coloration. This is a positive starch test.

Glucose tests. Put a small amount of glucose into the bottom of a test-tube. Half-fill the tube with water, and shake to dissolve all the glucose powder.

Benedict's test for glucose is carried out as follows:

Put 5·0 cm³ of Benedict's reagent into a test-tube and add about 8 drops of the prepared glucose solution to the reagent. Put the tube into a beaker of boiling water for 3–5 minutes. The mixed reagents change colour varying with the amount of glucose used; orange, red or brown colours result.

This is a positive test for a reducing sugar such as glucose.

Fehling's test for glucose is carried out as follows:

Put $2 \cdot 0$ cm^3 of Fehling's No. I solution into a test-tube; add the same quantity of Fehling's No. II. Add the same quantity of glucose solution. Put the tube into a beaker of boiling water and notice the change of colour to a deep orange-red after a few minutes. This is a positive test for a reducing sugar such as glucose.

Sucrose test requires that the sugar be broken up into glucose and fructose because Benedict's and Fehling's do not work with sucrose as above.

Prepare a sucrose solution and add Benedict's or Fehling's reagent. Try these two tests and notice a negative result. Use a fresh sucrose solution and add a few drops of dilute hydrochloric acid. Boil the solution for a minute or two. Repeat the glucose tests as previously and notice the colour change to show the presence of glucose.

Tests for glucose in the urine of diabetic people may be carried out as above, but generally tablets or paper strips soaked in chemicals are employed in medical laboratories.

EXERCISE 8.   TESTING FOR PROTEINS

REQUIREMENTS

Bunsen.burner, spatula, rack of pyrex test-tubes, watch glass, beaker 250 cm^3, tripod and gauze. Protein sample (egg albumen), Millon's reagent, Biuret reagent requirements. Concentrated nitric acid. Distilled water.

THEORY

Proteins are made up from long chains of amino-acids. They do not usually dissolve in water and form a gelatinous mass when heated in water.

They char to leave carbon on a spatula when heated strongly. They produce a typical "burning-flesh" smell.

EXPERIMENTS

Millon's test. Put some egg albumen into a test-tube. Add a few drops of Millon's reagent. A red colour is produced in the protein.

Biuret test. Make up a 1 % egg albumen solution in distilled water. Add 2 cm³ of the albumen solution to one tube and the same quantity of distilled water to a second tube. Add 2 cm³ of 10 % sodium hydroxide to each tube followed by 8–10 drops of 1 % copper sulphate solution. Compare the colours of each tube after about 5 minutes. A violet colour will be a positive protein test.

Xanthoproteic test. Put some egg white (egg albumen) flakes into a test-tube. Very carefully pour a few drops of concentrated nitric acid down the tube. A yellow colour is produced with proteins. (Be careful when disposing of the acid.)

EXERCISE 9. TESTING FOR FATS AND OILS

REQUIREMENTS

Bunsen burner, rack of pyrex test-tubes, oil sample (olive oil). Sudan III stain.

THEORY

Fats and oils do not dissolve in water. When they are shaken up with water a cloudy emulsion-like suspension of droplets is produced. These droplets drift to the water surface after a little time on standing.

EXPERIMENTS

Grease-spot test. Put some olive-oil drops into a test-tube. Add a few drops of ether or acetone (great care, no flames nor even hot gauze-inflammable ether) to the oil. Shake. Tip onto a clean filter-paper. Wave the paper to dry it and hold it up to the light to see if any grease spot is formed. The grease spot will suggest a fat or oil. (Of course, a spot will also be formed if the fat or oil is put directly on to the paper.)

Sudan III test. Put some oil into a test-tube. Cover the oil with

water. Shake and warm the tube over a flame. Add a little Sudan III
to the tube, shake and allow to settle. The fat droplets take up the red
stain.

EXERCISE 10. SWALLOWING

NO REQUIREMENTS

THEORY

Swallowing is an action involving several factors, such as the
presence of something in the mouth, movement of the larynx, cessation
of breathing.

EXPERIMENTS

(a) Swallow once, twice, three times. Is it possible to continue?

(b) Take a clean, absorbent tissue and dab it around the tongue and
cheeks. Try to swallow. Close the eyes and imagine drinking a sharp
lemon drink. Some action in the salivary glands can be felt.

(c) Hold the larynx whilst swallowing; notice the upward movement.

(d) Swallow and pay close attention to any noises in the ears. Why
are there noises as one swallows? Is one able to breathe *and* swallow?

(e) Some animals have no roof to the mouth; we do. What advan-
tages are there in having a separation between the nasal cavity and the
buccal cavity?

EXERCISE 11. EXAMINING THE ALIMENTARY CANAL AND ASSOCIATED ORGANS

REQUIREMENTS

A model of human abdomen. A freshly killed rabbit or rat.

THEORY

The alimentary canal varies in structure from animal to animal
depending upon the type of food eaten.

Man has an omnivorous diet and so has a varied selection of teeth for dealing with his varied diet of food. The stomach and intestines of man are also adapted to deal with this varied diet.

PROCEDURE

(a) Examine the model of the human abdomen and its contents, locate and identify the main organs. Write down the function of each organ as it is located.

(b) Make a close examination of a freshly killed dissected mammal such as a rat or a rabbit. (This animal will have been killed by a humane administration of anaesthetic.)

Using a seeker or the fingers move the abdominal organs about in order to locate underlying organs.

(c) Notice any points of difference from man.

The rat has no gall-bladder and its caecum is quite short. The rabbit has a very large caecum and a very long small intestine. Why?

EXERCISE 12. A DEMONSTRATION OF OSMOSIS ACROSS CELL MEMBRANES

REQUIREMENTS

Two beakers (250 cm³). 100 cm³ of 1% sucrose and 10% sucrose. Cellophane (Visking) tubing 2·5 cm diameter. 10 cm³ pipette. A weighing machine.

THEORY

All cells have their contents bounded by a plasma membrane. This membrane is described as semi-permeable because it allows some molecules to pass through it but holds others back.

Osmosis is a diffusion of water molecules across the membrane from an area of much water to an area of less water.

EXPERIMENT

It is possible to demonstrate the changes that occur in life by the following set of experiments. An "artificial cell" is created by filling a semi-permeable membrane tube with strong sugar solution. This sugar solution will represent the cytoplasmic contents.

(a) Cut up 15-cm lengths of the cellophane tubing. Five such tubes will be required.

Dampen the lengths of tubing and open them out. Tie up one end of each tube with a double knot.

(b) Fill up these "skins" with 10 cm³ of the following:

        Tube 1—tap water.

        Tube 2—1% sucrose solution.

        Tube 3—10% sucrose solution.

        Tube 4—tap water.

        Tube 5—10% sucrose.

(c) Squeeze all the air out of the tubes before making a double knot to enclose the fluids within the tubes.

(d) Prepare two beakers half-full with tap water in one case, the other half-full with 10% sucrose solution.

(e) Weigh all the tubes.

Place tubes 1, 2 and 3 into the beaker of tap water.

Place tubes 4 and 5 into the beaker of sugar solution.

Leave the tubes immersed in the beakers for about 2 hours (or overnight if more convenient).

(f) Remove each tube, wipe off excess fluid and weigh the tubes again. Record their new weights.

	Weight of tubes (grams)				
	1	2	3	4	5
Before					
After					

Results

Changes in the concentration of the fluid surrounding blood cells will cause the red cells to alter their appearance. An experiment to demonstrate osmosis and red blood cells follows later.

EXERCISE 13. ENZYME EXPERIMENT

The Digestion of Starch

REQUIREMENTS

Two beakers (250 cm^3), four test-tubes. 1% starch solution. Iodine for starch testing. Benedict's solution. A small amount of concentrated hydrochloric acid. A glass rod. A 5 cm^3 pipette. A thermometer. A source of heat.

THEORY

Starch is broken down (digested) by an amylase enzyme (ptyalin) found in saliva. Glucose is produced as a result of this digestion. Enzyme activity takes place at a definite pH value.

PROCEDURE

(a) Put 5 cm^3 of the starch solution into each of four tubes.

(b) Add to each tube the following:

Tube 1—5 cm^3 of water.

Tube 2—5 cm^3 of warm water which has been moved around the mouth for a minute or so.

Tube 3—5 cm^3 of warm water which has been moved around the mouth for a minute. Boil this mouth wash first before adding to the tube.

Tube 4—5 cm^3 of mouth wash to which has been added 1 drop of concentrated hydrochloric acid.

(c) Place all the tubes into a beaker of water warmed to body temperature.

(d) Add to each tube a few drops of iodine to give a blue starch reaction. Shake the tubes gently. Leave the tubes in the water bath for about 15 minutes and notice any colour changes.

(e) Whilst the digestion experiment is taking place, carry out a Benedict's test for glucose on the original starch solution used. Is any glucose present?

(f) When a definite colour change has taken place in tube 2 the experiment has finished.

(g) Test-tube 2 for the presence of glucose sugar.

Results

	Observations	Explanations
Tube 1		
Tube 2		
Tube 3		
Tube 4		

EXERCISE 14. ENZYME EXPERIMENT

The Digestion of a Protein

REQUIREMENTS

Two beakers (250 cm³), four test-tubes. Powdered fibrin or small pieces of hard egg white. 0·2% hydrochloric acid. Sodium hydroxide (dilute). 1% pepsin solution. A 2 cm³ pipette. A thermometer. A source of heat.

THEORY

Pepsin is a gastric protease which acts in an acid medium to break up proteins by rupturing some peptide linkages.

PROCEDURE

Prepare four tubes as follows:
(a) Put 1 cm³ of pepsin solution into each of three tubes.
(b) Add to each tube the following:
   Tube 1—2 cm³ of 0·2% hydrochloric acid.
   Tube 2—2 cm³ of sodium hydroxide.
   Tube 3—boil the pepsin, then add 0·2% hydrochloric acid.
(c) Put 1 cm³ of water in tube 4.

(d) Into all four tubes put a few grains of powdered fibrin or a small piece of egg white.

(e) Leave all four tubes in a beaker of water kept at body temperature. Examine the protein particles after 3 or 4 hours (longer if necessary).

Results

	Observations	Explanations
Tube 1		
Tube 2		
Tube 3		
Tube 4		

EXERCISE 15. ENZYME EXPERIMENT

The Digestion of Fat

REQUIREMENTS

Two test-tubes. A few drops of olive oil. A little alcohol. Phenol red indicator. Pancreatic lipase. 2% sodium carbonate. A beaker of water. A source of heat. A 2 cm³ pipette. A thermometer. A teat pipette.

THEORY

Fats are digested by lipases in slightly alkaline media. When fats are broken up fatty acids are liberated. These acids may be detected by a change in the colour of an indicator such as litmus.

PROCEDURE

(a) Three drops of olive oil are dissolved in 2 cm³ of alcohol by warming gently.

Add an equal volume of water and 5 drops of phenol red indicator.

(b) Add 3 cm³ of pancreatic lipase and make it just pink by careful addition of 2% sodium carbonate. Shake well.

(c) Put the tube in a beaker of water at body temperature for about 15 minutes.

(d) Make up a control tube using the same as above but with the lipase boiled and cooled.

(e) After 15 minutes examine the tubes for colour change.

Results

	Observations	Explanations
Experimental tube		
Control tube		

EXERCISE 16. ENZYME EXPERIMENT

The Coagulation of Milk by Rennin

REQUIREMENTS

Milk. Two beakers (250 cm³), four test-tubes, 5 cm³ pipette. Bottle of Rennet. 2% sodium carbonate. 2% acetic acid. A source of heat. A thermometer.

THEORY

Rennin is a proteolytic enzyme found in the gastric lining of suckling animals. It converts caseinogen into insoluble casein (easier to digest).

PROCEDURE

Set up four tubes as follows:
Tube 1—3 cm³ of milk with 1 cm³ of Rennin.
Tube 2—3 cm³ of milk with 1 cm³ of boiled Rennin.

Tube 3—3 cm³ of milk with 1 cm³ of Rennin and 2 drops of 2%
   acetic acid.
Tube 4—3 cm³ of milk with 1 cm³ of Rennin and 1 cm³ of 2%
   sodium carbonate.
   Examine the tubes for the formation of a milk clot. Leave the tubes
in a beaker of water at body temperature.

Results

	Observations	Explanations
Tube 1		
Tube 2		
Tube 3		
Tube 4		

EXERCISE 17.   EXAMINING THE RESPIRATORY ORGANS

REQUIREMENTS

Anatomical model of the thorax. A freshly killed small mammal
Dissection instruments. Trachea and lungs of large mammal from
butcher (or abbatoir).

THEORY

Air is conducted to the air sacs by way of the nasal cavity, trachea,
bronchi and bronchioles. These organs provide the body with oxygen
which is exchanged for carbon dioxide within the alveoli.

PROCEDURE

(a) Examine an anatomical model and/or wall chart of the rib-cage
and its contents. Locate the main parts of the respiratory system. How
many lobes are there on each lung? Is each lung identical?

(b) Handle a pair of lungs and attached trachea obtained from a cow or sheep. Notice the texture of the pink spongy lungs. What colour may they be if a person smokes cigarettes? Have a close look at the cartilage hoops supporting the trachea.

Put a rubber tube down the trachea; attempt to inflate the lungs.

(c) Examine a dissected mammal thorax. Notice the close association of the heart and lungs. Note the texture and structure of the muscular diaphragm.

EXERCISE 18. INSPIRATION AND EXPIRATION

REQUIREMENTS

Bell-jar model of the rib cage. Flasks containing limewater fitted with rubber tubing. Dry cobalt chloride paper. Stethoscope.

THEORY

As carbon dioxide builds up in the blood-stream a nervous message brings about a diaphragm movement and air is inspired. The carbon dioxide is expired with water vapour.

PROCEDURE

(a) Operate the model rib-cage and diaphragm. By pulling down the rubber "diaphragm" air enters the balloon "lungs". Push the rubber sheet upwards and air is forced out. Compare this with the rib and diaphragm movements in the human body. How many times per minute do we ventilate our lungs?

(b) If a stethoscope is available listen to the breathing sounds of a partner. Listen from the front of the rib-cage and from the back to see if there are any different sounds when the breathing is deep.

(c) Breathe through the limewater as shown in Fig. 106 Compare the colour of the limewater in both flasks after about 5 minutes. Limewater (calcium hydroxide solution) goes cloudy when carbon dioxide is bubbled through it.

(d) Breathe heavily onto a piece of dry cobalt chloride paper. It is sensitive to water. What colour does it change to?

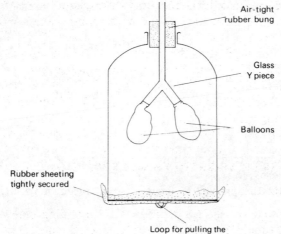

Air-tight
rubber bung

Glass
Y piece

Balloons

Rubber sheeting
tightly secured

Loop for pulling the
rubber sheet downwards

(a)

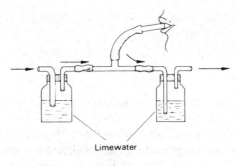

Limewater

(b)

Fig. 106. (a) The bell-jar experiment. (b) Breathing through limewater.

EXERCISE 19.   ENZYMES WITHIN THE CELL

REQUIREMENTS

Hydrogen peroxide. Blood and muscle tissue from freshly killed mammal (mouse). Watch glass.

THEORY

Oxygen and glucose are "treated" by many enzymes within the cell (intra-cellular enzymes). The result of this "treatment" is the liberation of energy for living processes.

All cells contain an enzyme called catalase which destroys hydrogen peroxide if it is formed during intra-cellular activity.

EXPERIMENT

(a) Put some fresh blood into a watch glass. Add a few drops of hydrogen peroxide.

(b) Repeat the procedure using a small piece of freshly chopped muscle.

Result

	Observation	Explanation
Experiment (a)		
Experiment (b)		

EXERCISE 20. MAKING A BLOOD SMEAR

REQUIREMENTS

Microscope, lamp, glass slides. Leishman's stain. Distilled water of pH 6·8 (see exercise 32). A source of fresh blood.

THEORY

Making a blood film is frequently necessary in hospital laboratories. To make a good blood film it is necessary to have clean glassware and a smooth-edged slide for spreading the blood.

PROCEDURE

(a) Blood may be taken from a finger or thumb by making a swift puncture with a sterile lancet. The first drop of blood is wiped away with clean cotton wool. The second drop is put near to the end of a clean microscope slide. The area must not be squeezed as this will not produce a normal blood drop. The skin area is wiped clean with cotton wool dipped in surgical spirit or alcohol before and after the operation.

A blood lancet used for this must be thrown away after use. Such a lancet must *not* be used by any other person.

The above method of obtaining blood, it has been argued, should be discouraged because of the dangers of a virus disease hepatitis.

(b) Blood can be obtained from a freshly killed mammal or from some commercial source.

(c) Spreading the blood across the microscope slide must be done swiftly in the manner shown in Fig. 107.

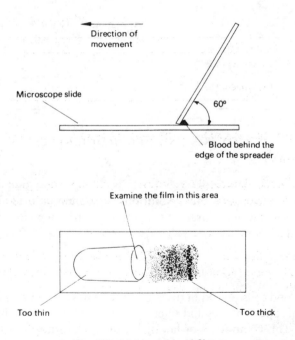

FIG. 107. Making a blood film.

The film of blood produced must not be too thick.

(d) Staining the blood film to show up the cells can be carried out as follows:

(i) Dry the smear or film by waving it in the air for a short time.

(ii) Cover the film with Leishmán's.stain. Leave it on the film for 1 or 2 minutes.

(iii) Add some distilled water (pH 6·8) to the stain until the whole film is flooded with the diluted stain. Leave for about 10 minutes until the film becomes pinkish.

(iv) Drain the slide and allow it to dry at room temperature. Wipe the back of the slide to dry and examine the film under the low power of the microscope.

Result

	Appearance and colour
Erythrocytes	
Leucocytes	
Platelets	

EXERCISE 21. OSMOSIS AND RED BLOOD CELLS (HAEMOLYSIS)

REQUIREMENTS

Test-tubes, test-tube rack. Solutions of sodium chloride 0·9% and 10%. Distilled water. Three teat pipettes. Microscope and microscope slides and a lamp.

THEORY

Red blood cells placed in distilled water will swell up and burst. The cell membrane ruptures and haemoglobin is released into the solution (laking). This haemolysis or laking is caused by osmosis. Water passes across the semi-permeable membrane of the cell. The cell membrane is more fragile in some blood diseases and will be more easily ruptured by changes in fluid concentration.

PROCEDURE

(a) Prepare three test-tubes as follows:
Tube 1—half full with 10% sodium chloride solution.
Tube 2—half full with 0·9% sodium chloride solution.
Tube 3—half full with distilled water.
To each tube add 5 drops of blood.
(b) Leave the tubes in a test-tube rack after inverting the tubes to mix.
(c) After about 30 minutes look at the tubes closely; a lamp behind the tubes is helpful. Record what appears to have happened.
(d) Take a small sample from each tube and examine them under the microscope. Describe what appears to have happened. Human blood cells begin to haemolyse in salt solution of about 0·5%. Saline is described as *"normal saline"* (0·9% salt solution) because the red blood cells show no change in 0·9% salt solution. Normal saline is said to be isotonic with blood. There is no water uptake.

Results

	Observations	Explanations
Tube 1		
Tube 2		
Tube 3		

EXERCISE 22. EXAMINING THE HEART

REQUIREMENTS

Fresh sheep's heart. Dissecting instruments. Stethoscope. Heart models. Watch or clock with second hand.

THEORY

The heart is a four-chambered muscular pump. The pumping action of the heart can be examined by means of a stethoscope and by examining the pulse and blood pressure.

PROCEDURE

(a) *Examining the sheep's heart*

Lay the heart on a board with the ventral surface facing uppermost. Identify the main vessels leaving the heart and the vessels supplying the heart itself (coronaries). Use a model and a labelled heart diagram, if available, to help in the identification. Make an incision in the thick ventricle muscle in a direction parallel to the table. Continue this incision around the apex of the heart. Finally, make a cut across the heart just beneath the atria. A "window" of ventricle muscle should now be removed to expose the interior of the ventricle cavities. Locate the valves between atria and ventricles and the chords which hold the valves in position.

Open up the atria and poke a seeker through the great blood vessels to note their openings into the atria.

(b) *Listening to heart sounds*

It is possible to hear the heart action by putting the ear to the chest wall of a person. Using a stethoscope is a more convenient method. Two noises can be heard: "lub–dup".

The first noise is associated with the ventricles squeezing blood out of the heart. The valves between atria and ventricles close at this time. The second noise is associated with the blood in the atria being forced down into the ventricles. The aortic and pulmonary valves close at this time.

(c) *Taking the pulse*

Rest the forearm on the table. Rest three fingers along the thumb side of the forearm and apply a gentle pressure until the pulsing of the artery beneath the skin is felt.

Carry out the following experiments:

(i) How many times does the pulse beat in a minute?

(ii) Breathe in and out deeply about 10 times and note if there is any change in the pulse rate.

(iii) Press the finger resting on the artery, nearest the heart, until the pulse is no longer felt with the fingers nearest the hand. The pressure required to stop the pulse in this way gives one some idea of the feel of normal blood pressure.

(iv) One way of determining a person's fitness is to check that he is able to carry out some energetic task without the heart undergoing too much stress. Try the following experiment with a partner.

First, take the resting pulse rate per minute.

Second, ask the partner to stand up and down onto a chair five times, bringing both feet onto the chair for each count. Keep the chair steady. Immediately he has completed five "step-ups" take his pulse rate per minute.

Third, let the partner sit down for 5 minutes and take the pulse rate per minute again.

If five "step-ups" is insufficient to produce a noticeable pulse change then increase the exercise to eight or ten "step-ups".

Results

Record the results as below:

Pulse rates per minute

At rest	After deep breathing	After exercise	At rest after exercise

Compare the results obtained with others.

EXERCISE 23. ANALYSING URINE FOR ABNORMAL CONSTITUENTS

REQUIREMENTS

Trade literature supplied with testing strips and tablets, obtainable from chemist shops.

Urine or "mock urine" with abnormal constituents added for testing purposes. Urine or "mock urine" without abnormal constituents. Reagent strips to demonstrate presence in urine of protein, blood, glucose, ketones, phenylpyruvic acid, bile pigments.

THEORY

The presence of abnormal constituents in the urine is quickly tested for by hospital laboratory workers using paper strips soaked in chemical testing reagents.

The presence of these abnormal constituents in urine is or may be an indicator of disease.

PROCEDURE

Prepare five tubes of "abnormal urine" and test each using the reagent strips as indicated on the manufacturer's literature. Prepare other tubes without abnormal constituents. Record the possible cause of the presence of these abnormal constituents.

Results

Abnormal constituents	Possible causes
Protein (e.g. albumen)	
Blood	
Glucose	
Ketones (e.g. acetone)	
Phenylpyruvic acid	
Bile pigments	

EXERCISE 24. EXAMINING BODY TEMPERATURE

REQUIREMENTS

Clinical thermometer. Thermistor thermometer (from physics laboratory or easily constructed).

THEORY

Man is able to maintain a constant body temperature but this really only applies to the "core" of the body, the thorax and abdomen contents as well as those of the skull. The extremities of the body may be well below this internal temperature.

In fever this temperature regulation ability is upset.

PROCEDURE

(a) Examine a clinical thermometer and note that it must be shaken sharply to return the mercury to the bulb.

(b) Clean the thermometer with soap and water, dry and put into the mouth beneath the tongue. Keep the mouth firmly closed. Leave the thermometer in the mouth for $\frac{1}{2}$ minute and record the temperature. Repeat and record the temperature after 1 minute under the tongue.

(c) If possible, record the temperature of the armpit (keep the thermometer trapped in the armpit for 1 minute but with care not to break the glass).

(d) Carry out some tough exercise whilst dressed in overcoat and scarf. Step up and down onto a chair five to ten times and record the temperature afterwards to see if there is any rise. Try longer periods of exercise if necessary.

(e) Check that the electronic thermometer (thermistor) is working. Take temperature readings from the skin at various points. Start by taking readings from the uncovered arm. Take readings at different points along the length of the limb.

Temperature recordings

	Shoulder	Arm-pit	On biceps	Elbow	Mid-forearm	Wrist	Finger-tips
In warm air							
In cold air							

EXERCISE 25. A STUDY OF THE SKIN

REQUIREMENTS

Two plastic bowls. Metal knitting needle (or similar). A source of heat. Cotton wool. Plasticine. Pins. Ruler. Microscope slide mirror.

THEORY

The skin is an important sense organ. Much of our knowledge of the environment comes to us by way of the skin.

PROCEDURE

Discrimination

(a) Put two pins in a piece of plasticine. Adjust the distance between the points, starting with a distance of 1 mm. Increase by 1 mm for each experiment.

(b) Ask a partner to close his eyes. Touch his skin at various points on the body and ask him to discriminate between one or two points. (Do not allow the partner to see the double and single pins before the experiment.)

(c) Touch the arms, legs, shoulders, face, tongue, finger-tips and elsewhere as possible.

Record the results in the table on p. 273.

Touch sensation

(a) Fix some cotton wool to the end of a matchstick of knitting needle. Do not let your partner see this.

Discrimination test results (+ . —)

Distances apart	Face	Tongue	Neck	Shoulder	Forearm	Finger-tips	etc.
1 mm							
2 mm							
3 mm							
4 mm							
etc.							

(b) Touch the skin lightly on various parts of the body. Ask the partner to say when he can feel something and to describe what sensation he feels. Record your results from touching the face, lips, hands, leg, etc. Determine which area is the most sensitive.

Skin adaptation

(a) Prepare a bowl of hot water and a bowl of cold water. Ask your partner to plunge one hand into the cold water, the other into the hot water. Leave the hands in the bowls for a few minutes.

(b) Tell your partner to move his hands into two new bowls of water or change the bowls around so that the hand adapted to hot water is now in cold water. Is it easy to tell the difference?

Repeat the experiment and plunge the hands into tepid water after adapting to hot and cold water. What sensations are felt?

Skin capillaries

Try the following two experiments:

(a) Press gently on the nail until the nail bed goes white.

(b) Press a glass microscope slide onto the lip until the lip goes white. Examine the blood return (by looking in a mirror) as the pressure is eased.

Skin response to injury

Scratch the skin of the forearm with a blunt instrument, like a kintting needle or pencil.

Notice the red weal produced—a mild inflammation response.

EXERCISE 26. STUDYING REFLEX ACTION

REQUIREMENTS

A mirror. A ruler bound up in a little bandage.

THEORY

A spinal reflex is an involuntary response to a stimulus involving nervous pathways confined to the spinal cord.

PROCEDURE

(a) Cover both eyes with the palms of the hands. Sit before a mirror in this fashion for 5 minutes. Remove the hands from the eyes and look closely at the pupils in the mirror. If you notice nothing repeat the procedure.

The pupillary reflex causes the pupil to dilate (open) when little light is present. On exposure to bright light the pupils close. One way of checking upon the state of an apparently unconscious person is to open the eyelid to look for a pupillary reflex.

(b) Sit a partner on a chair with his legs crossed. Locate the tendon beneath the knee-cap (patella). Tap this tendon firmly with the ruler. A sudden stretching of the quadriceps muscle of the thigh causes it to contract and an involuntary "knee-jerk" reflex is produced.

Try a similar experiment by tapping the Achilles tendon behind the ankle.

Tickle the sole of somebody's foot for a moment or two and notice the result. This works particularly well with a baby.

EXERCISE 27.   A STUDY OF THE EYE

REQUIREMENTS

Ox or sheep eye for dissection. Dissection instruments. Blind spot diagram. Coloured pencils. Watch glass. Eye model and labelled charts. Ruler. Mirror. Sheet of white paper. Coloured papers.

THEORY

The eyes are able to focus on near and distant objects, to perceive colour, and to adjust to changes of illumination by altering the pupil size.

PROCEDURE

Examining the eye by dissection

Remove all the fat tissue surrounding the eyeball. Leave the optic nerve in position. Examine the cornea; locate the iris and pupil. Cut around the edge of the cornea where it joins the outer tough sclera. Remove the front of the eye and put it into a watch glass. Locate the aqueous humour, the lens, iris and ciliary processes. Remove the lens and cut it in half to move the internal structure. Remove the vitreous humour from the main body of the eye in order to expose the retina. Examine the retina closely to look for the exit of the optic nerve (the blind spot). Notice that the optic nerve does not leave the centre of the posterior surface.

Locating the blind spot

Close the left eye. Hold the book about 30 cm away and look

FIG. 108. Locating the blind spot.

at the cross with the right eye. Bring the book closer. At about 25 cm the circle on the right will disappear.

Visual accommodation

The near point of vision is the distance from the eye of the nearest object that can be focused clearly. Cover one eye with the hand. Hold a pencil at arm's length and focus on it as it is brought closer to the eye. Stop when the pencil can no longer be seen sharply. Measure this distance.

Repeat this procedure for the other eye. Hold a pencil close to the nose of a partner. Ask him to look away at a distant object. Then tell him to quickly focus on the pencil.

Notice as the eyes converge on the pencil that the pupils become smaller.

Pupillary reflex reaction to light

Cover one eye with the hand for 1 minute. Look into a mirror and then remove the hand from the eye. Watch the pupil. A contraction (closing) of the pupil is followed by a slight dilation.

Colour vision

Sit at a desk and fix your eyes on an object straight ahead. Pick up a coloured crayon (without knowing its colour) and hold it at arm's length behind the head and out of vision. Move the crayon forwards slowly and stop when it comes into the visual field. Keep looking straight ahead. Is it possible to see what colour the pencil is? Cones are not very plentiful at the outer edges of the retina. Some colours will be seen more easily than others at the edge of the visual field.

After-images

Stare hard for $\frac{1}{2}$ minute at a circle of red paper. Transfer the eyes to a white paper. What colour is seen? Close the eyes and what image is seen?

Repeat this procedure with other colours.

An after-image is known as a negative after-image if the colour is complementary to the original. A positive after-image is in the same colour as the original. Stare for $\frac{1}{2}$ minute at a bright white light and then look at a sheet of white paper. What image is seen?

EXERCISE 28.   A STUDY OF THE EAR AND HEARING

REQUIREMENTS

Model of the ear. Labelled chart. Watch. Tuning fork.

THEORY

The ear is an organ of hearing and balance. Air vibrations are con-ducted to the organ of hearing by way of the eardrum and through the bones of the skull.

PROCEDURE

(a) Examine the model of the ear and identify the main parts associated with hearing and balance.

(b) Hearing may be tested as follows:

The watch test involves the faint ticking of a watch. Ask a partner to cover up one ear. Bring the watch up towards the back of his head and stop when he can hear it.

Measure this distance.

Repeat the observation for the other ear.

The tuning-fork hearing test involves a vibrating tuning fork sending vibrations through the air and through the skull bones.

   (i) Strike the fork on the hand and bring it up towards one ear, the other ear being covered. Measure the distance when the vibrations are first heard. Repeat this test for the other ear.

  (ii) Strike the tuning fork against the hand and then put the handle on the mastoid bone behind the ear. Compare this bone-conduction hearing with that of air-conduction hearing.

 (iii) Place the vibrating fork handle onto the bone of the middle of the forehead. Which ear receives the clearest vibration?

(c) Balance-testing experiments may be carried out if a rotating chair is available.

   (i) Rotate a person in such a chair about 10 times in 20 seconds, clockwise. Ask him to focus on some distant object as he is rotated. Examine his eyes for any continued movement after the rotations have stopped.

   (ii) Try to stand on one leg blind-folded and with normal vision. What difference does blind-folding make?

EXERCISE 29.   TASTE AND SMELL

REQUIREMENTS

A selection of test solutions. 5% sodium chloride solution. 2% citric acid solution. 1% quinine disulphate. 5% sucrose solution. An apple, an onion, a banana, sugar. A watch. A selection of odiferous substances such as ammonia, camphor, acetic acid. Clean paper tissues. Cotton wool. Watch glasses.

THEORY

Taste and smell are closely related. Interference with sense of smell can affect tasting.

PROCEDURE

Taste

(a) Dry the tongue with a clean paper tissue. Put a little sugar onto the tongue. Can it be tasted immediately? How long did it lay there before giving a taste?

(b) Draw an outline diagram of the tongue. Ask a partner to dry his tongue with a tissue. Dip a swab of cotton wool into one of the test solutions. Move this swab over your partner's tongue and later ask him where he tasted the test solution most distinctly. (Do not tell him the solutions you are using until after the experiment.) Wash out the mouth, dry the tongue and repeat the procedure using the other test solutions. Draw in the taste areas on the outline tongue diagram.

(c) Ask your partner to close his eyes, pinch his nostrils together and stick his tongue out. Have some previously prepared small portions of apple, onion and banana.

Put a sample of each onto the tongue of your partner. Look at your watch and see how long it takes before he is able to tell you what food he has. Attempt this for each food mentioned. Without smelling it is sometimes difficult to be sure of taste.

Repeat this experiment but allow your partner to swallow the food straight away. Is there any difference in the time taken for recognition?

Smell

(a) Ask your partner to close his eyes. Bring different smelly substances into the room on watch glasses. Place the dish 2 m in front of your partner and see how long it is before he can identify or smell some odour.

Move the odour nearer if the time exceeds 5 minutes.

(b) Repeat the above experiment but unknown to your partner bring in a mixture of various smelly substances and see which one he recognizes or smells first. Or does the mixture neutralize any particular odour?

EXERCISE 30. THE DEVELOPING EMBRYO

REQUIREMENTS

Saline solution (mammal saline 0·9 % sodium chloride). Deep-sided dish, like a crystallizing dish. Half a dozen fertile chicken eggs. An incubator or warm place of fairly constant temperature [38°C (100°F) and 50 % humidity]. Dissecting instruments. A dissecting lens or dissecting microscope. A lamp. Plasticine.

THEORY

The developing embryo of man obtains nutrients from the blood of the mother. The developing embryo of the chicken obtains its nutrients from the stored yolk.

PROCEDURE

A developing embryo may most easily be seen in the fertile chicken egg. Place half a dozen fertile chicken eggs into an incubator or suitable area for incubation.

Remove an egg every 4 days for examination. Hold the egg with the pointed end to the left. Tap the central area of the egg shell to break it. Remove a window of shell with scissors to reveal the developing embryo. Immerse the egg in a warm saline solution and set up a lens or microscope to observe the embryo. The head of the embryo will be facing the observer in this situation. The egg may be supported in the salt solution on a ring of plasticine.

EXERCISE 31.   EXAMINING SPERMATOZOA

REQUIREMENTS

A freshly killed earthworm. Dissecting instruments. Microscope and lamp. A microscope slide. Saline solution (0·6%).

THEORY

Spermatozoa may be obtained from the spermathecae of the earthworm. They have a limited life when removed from the body.

PROCEDURE

A teacher or demonstrator will prepare a smear of spermatozoa from an earthworm. Examine the spermatozoa under the microscope. If conditions permit, each student could prepare a seminal smear in the manner demonstrated by the instructor. Look out for evidence of a parasite called *Monocystis* which lives in the seminal vesicles of most earthworms.

EXERCISE 32.   PREPARING A SOLUTION OF pH 6·8
(for Exercise 20)

REQUIREMENTS

Sodium dihydrogen phosphate ($NaH_2PO_4 \cdot 2H_2O$). Disodium hydrogen phosphate (Na₂HPO₄). Chemical balance. Beaker (1000 cm³). Measuring cylinder (1000 cm³). Stirring rod. De-ionized water.

Solutions of a particular pH are required for many experiments in human physiology.

Make up a solution of sodium dihydrogen phosphate—31·2 g made up to a litre with de-ionized water. Make up a solution of disodium hydrogen phosphate—28·39 g made up to a litre with de-ionized water. In order to make up a solution of pH 6·8 add together the amounts indicated below. Make up to 1000 cm³ with de-ionized water.

$$25 \cdot 5 \text{ cm}^3 \text{ (NaH}_2\text{PO}_4) + 24 \cdot 5 \text{ cm}^3 \text{ (Na}_2\text{HPO}_4 \cdot 2\text{H}_2\text{O)}, \text{ pH } 6 \cdot 8.$$

The pH of this solution can be checked by using a battery or a mains pH meter. The instructions for its use are usually included with the apparatus.

EXERCISE 33. GROWING BACTERIA (AND OTHER MICRO-ORGANISMS)

Six nutrient agar plates (disposable) made up by the teacher or technician. The Oxoid *Microbiology for Schools* booklet of useful experiments. Incubator or warm place.

Micro-organisms grow readily on a nutrient agar. Agar is a polysaccharide extracted from seaweed which melts at about 95°C (203°F) and solidifies at 42°C (108°F). The nutrient is a meat extract. Prior to the experiment all materials must be sterile.

PROCEDURE

Expose a number of nutrient agar plates to possible sources of contamination:

(i) Touch with the fingers.
(ii) Leave exposed to the air for a short while.
(iii) Cough onto the plate.
(iv) Rub a piece of clothing across the plate.

Try other sources of contamination in your place of study. Incubate the plates for 2 or 3 days.

RESULTS

Observe the growth of micro-organisms with a lens or microscope. After the experiment all the plates are destroyed by heat.

Appendix

Addresses of Scientific Suppliers

Teachers are strongly advised to obtain the catalogues from some or all of the following well-known equipment suppliers.

Baird and Tatlock—Freshwater Road, Chadwell Heath, Essex. (01·590·7700.)

Brain Sciences Information Project—Millbank Tower, London, SW1P 4QS. (01·834·6611.)

Campden Instruments—186 Campden Hill Road, London, W8. (01·727·3437.)

Colne Instruments Co. Ltd.—51 Lion Road, Twickenham, Middlesex. (01·892·7444.)

Gallenkamp—P.O. Box 290, Technico House, Christopher Street, London, EC2. (01·247·3211.)

Gerrard and Haig Ltd.—"Gerrard House", Worthing Road, East Preston, nr. Little-hampton, Sussex. (Rustington 090·624151.)

Griffen and George—Ealing Road, Alperton, Wembley, Middlesex, HAO 1HJ. (01·998·7711.)

Harris, Philip Ltd.—63 Ludgate Hill, Birmingham B3 1DJ. (021·236·4041.)

Oxoid Laboratory Products—Oxoid Ltd., Southwark Bridge Road, London, SE1 9HF. (01·928·4515.)

Palmer,C. R. (London) Ltd.—Myographic Works, Effra Road, Brixton, London SW2. (01·733·2173.)

Some Other Useful Addresses

Department of Health and Social Security
Alexander Fleming House, Elephant and Castle, London, SE1.
Inquiries to Information Division. (01·407·5522 Ext. 6555.)
Publications: *Notes on the National Health Service.*

Central Film Library
Government Building, Bromyard Avenue, Acton, London, W3. (01·743·5555.)
Publications: film catalogues.

General Dental Council
37 Wimpole Street, London, W1. (01·486·2171.)
Publications: films, posters, booklets about teeth.

The Institute of Biology
41 Queen's Gate, London, SW7. (01·589·9076.)
Publications: *Careers in Biology. Studies in Biology.*

National Council of Social Service
26 Bedford Square, London, WC1B 3HU. (01·636·4066.)
Publications: *Community Studies.*

The National Organisation for Audio–Visual Aids
33 Queen Anne Street, London, W1. (01·636·5742.)
Publications: *Advisory Service for Teachers.* Catalogues of films and film-strips.

The National Audio–Visual Aids Library
Paxton Place, Gipsy Road, London, SE27.
Publications: films and film-strips—catalogues.

National Society for Clean Air
134 North Street, Brighton, BN1 1RG. (Brighton 26313.)
Publications: books and films.

Oxfam Education
274 Banbury Road, Oxford, OX2 7DZ. (Oxford 56777.)
Publications: posters, books, project suggestions.

Bird's Eye Foods Ltd.
Educational Service, Station Avenue, Walton-on-Thames, Surrey.
Publications: wall charts and books on foods and diets.

Cheese Bureau
40 Berkeley Square, London, W1X 6AD. (01·499·1985.)
Publications: booklets, wall charts on protein foods.

Energen Foods Co. Ltd.
Birling Road, Ashford, Kent. (Ashford 23411.)
Publications: slimming and dieting booklets.

Nutrition Information Centre (Beecham Foods)
Beecham House, Great West Road, Brentford, Middx. (01·560·5151.)
Publications: folder charts on balanced diets.

(All the above addresses are liable to change)

Bibliography

Laboratory Information

The teacher or technician preparing solutions and reagents for any experiment with which they are not familiar are recommended to make reference to the publications marked*.

Mostly for the Teacher

AYKROYD, W.R., *Food for Man*, Pergamon Press, 1964.
*BAKER, F.J., SILVERTON, R.E. and LUCKCOCK, E.D., *Introduction to Medical Laboratory Technology*, Butterworths, 1966.
*BROWN, G.D. and CREEDY, J., *Experimental Biology Manual*, Heinemann, 1970.
BUTLER, J.A.V., *The Life Process*, George Allen & Unwin, 1970.
CLEGG, A.C. and CLEGG, P.C., *A Biology of the Mammal*, Heinemann, 1970.
CLEGG, A.C. and CLEGG, P.C., *Man Against Disease*, Heinemann, 1973.
FISHER, R.B. and CHRISTIE, G.A., *A Dictionary of Drugs*, Paladin, 1971.
GOMEZ, J., *A Dictionary of Symptoms*, Paladin, 1967.
*HALE, L.J., *Biological Laboratory Data*, Methuen, 1965.
HURRY, S., *Microstructure of Cells*, Murray, 1968.
JONES, B.R., *Pharmacology for Student and Pupil Nurses*, Heinemann, 1971.
LAURIE, P., *Drugs*, Penguin, 1967.
MATTHEWS, B.F., *Chemical Exchanges in Man*, Oliver & Boyd, 1967.
MCNAUGHT, A.B. and CALLENDER, R., *Nurses' Illustrated Physiology*, Livingstone, 1964.
NOSSAL, G.J.V., *Antibodies and Immunity*, Penguin, 1969.
*OTTO, TOWLE and CRIDER, *Biology Investigations*, Holt, Rinehart, Winston, 1965.
Oxford Biology Readers. Ed. J.J. HEAD and O.E. LOWENSTEIN, O.U.P., 1970.
PAUL, J., *Cell Biology*, Heinemann, 1970.
*PEACOCK, H.A., *Elementary Microtechnique*, Revised by S. Bradburg, Arnold, 1973.
PRIEST, M.A., *Modern Textbook of Personal and Communal Health for Nurses*, Heinemann, 1966.
ROSE, S., *The Chemistry of Life*, Penguin, 1966.
*SIROCKIN, G. and CULLIMORE, S., *Practical Microbiology*, McGraw-Hill, 1969.
WILLMOTT, P., *Consumer's Guide to the British Social Services*, Penguin, 1967.
WINTON and BAYLIS, *Human Physiology*, Churchill, 1968.

286   *Bibliography*

*Turtox Products, 8200 South Hoyne Avenue, Chicago, Illinois 60620, supply very useful leaflets for technician and teacher.

Useful for Student or Teacher

ABERCROMBIE, M., HICKMAN, C.J. and JOHNSON, H.L., *A Dictionary of Biology*, Penguin, 1973.
ALEXANDER, P., *Atomic Radiation and Life*, Penguin, 1959.
BARNETT, A., *The Human Species*, Penguin, 1957.
BROCKINGTON, F., *World Health*, Penguin, 1959.
BURNETT, F.M., *Viruses and Man*, Penguin, 1953.
HAY, D., *Human Populations* (Biology Topic Books), Penguin, 1972.
KALMUS, H., *Genetics*, Penguin, 1952.
McKUSICK, V.A., *Human Genetics*, Prentice-Hall, 1969.
MORRIS, D., *The Naked Ape*, Corgi Books.
PROBERT, A.J., *Parasites* (Biology Topic Books), Penguin, 1972.
ROMER, A.S., *Man and the Vertebrates* (II), Penguin, 1954.
SKURNIK, L.S. and GEORGE, F., *Psychology for Everyman*, Pelican, 1967.
STONEMAN, C.F., *Space Biology* (Biology Topic Books), Penguin, 1972.
YUDKIN, J., *This Slimming Business*, Penguin, 1958.

These reading lists represent source material for the more advanced pupil or student, and for those carrying out projects or essays on selected topics.

Glossary and Index

The glossary together with index will help the student do a little 'self-teaching'. The glossary may be used as an aid to 'self-testing' when the time for examination revision comes around.

287

Arteries. Blood vessels which carry blood away from the heart. They have muscular walls.   112

Ascorbic acid. Vitamin C soluble in water. Deficiency causes scurvy and difficult wound healing. Found in fruit juices.   57

Astigmatism. An eye condition in which the cornea has an unequal curvature in one or more directions.   166

Atlas. First vertebra of the vertebral column, next to the skull. Allows nodding of the head.   30

Auditory meatus. Tube leading from outside to ear drum. It is lined with whiskers and wax glands.   161

Autonomic nervous system. Section of nervous system not under the control of the "will". Branches $<$ *Sympathetic system* / *Parasympathetic system*   150

Axis. Second vertebra of vertebral column. The atlas-axis joint allows head to rotate.   30

Bacteria. Unicellular, microscopic organisms having various shapes. Many bacteria are pathogenic. Cocci—spherical bacteria. Bacilli—rod-shaped bacteria. Spirochaetes—spiral-shaped bacteria.   228

Balance. Ability associated with the semicircular canals in the inner ear and the cerebellum of the brain   165

Benedict's test. A test to determine the presence of glucose.   48, 251

Beri-beri. Diseased condition resulting from severe vitamin B deficiency.   56

Biceps femoris. Muscle behind the thigh, used to bend the leg.   39

Bile duct (common). Duct carrying bile juice from the gall bladder to the duodenum.   72

Bile juice. A secretion of the liver cells, pigmented by the products of haemoglobin decomposition. Used to emulsify fats in digestion.   75, 79

Bile pigments. Bilirubin—reddish. Biliverdin—brownish-green. Produced from haemoglobin breakdown in the liver. Present in the blood and urine of jaundiced patients.   79

Bladder. Muscular reservoir containing urine; opening to the exterior by way of the urethra.   125

Blind spot. The area on the retina where the nerve fibres pass to the brain by way of the optic nerve.   156

Blood. Fluid transporting food and oxygen to and from areas of the body. Composed of cells floating in a fluid plasma.   94

Bone. A hard connective tissue, built up from cartilage by the laying down of calcium salts (mainly phosphates) during development.   16, 21

Bowman's capsule. Capsule of the Malpighian body in which the glomerulus rests. Blood filtered across the capsule wall in the first stage of urine formation.   121

Brain. "Operation centre" of the central nervous system consisting of nerve cells and fibres.   136

Breathing. The inspiration and expiration of air by means of the muscular contractions and relaxations of the diaphragm and ribs.   86

Bronchi. The main tubes leading from the trachea to the lungs.   85

Caecum. A blind ending of the large intestine from which the appendix arises.   76

290   *Glossary and Index*

Triceps. The extensor muscle on the arm. 40
Trunk cavity contents. 18
Trypsin. A proteolytic enzyme in the pancreatic juice. 63
Tympanic membrane. The ear drum. 162

Ulna. The long bone on the inner side of the forearm. Bigger than the radius. 28
Umbilical cord. The cord which connects the foetus with the placenta and through which blood vessels run. 191
Urea. The main nitrogenous excretory product in urine. 123
Ureters. Tubes conducting urine from the kidneys to the bladder. 119
Urethra. Tube leading urine from the bladder to the outside. 119
Urinary system. 119
Urine. Fluid excretory product manufactures in the kidneys. Contains mainly urea and water. 123
Uriniferous tubules (nephrons). Tubules in the kidney which produce urine from the blood. 121
Uterus. The womb. A muscular organ about 3 in. in length and 2 in. wide. It changes in size as the foetus grows. 184
Urula. Part of the soft palate. 66, 71

Vagina. Sheath-like passage leading from the vulva to the uterus. 186
Vagus nerve. The nerve which travels to many thoracic organs. It slows the heartbeat. 142, 152
Vas deferens. Tubes leading from testes to penis. 181
Vasopressin (pitressin). Hormonal extract from posterior lobe of pituitary. 173
Veins. Vessels carrying blood back to the heart. 112
Vena cava. The main vein carrying blood into the heart. 107
Ventilation, methods of. 212
Vertebral column. Bony column running up the back; made up of 33 bones. 29
Villi. Finger-like projections in the small intestine across which digested food products diffuse. 74
Vitamins. Essential amino-acids found in foods. 54
Vitreous humour. Thick fluid behind the lens and in front of the retina. 157
Voluntary action. An action involving the "higher" centres of the brain. 139
Vulva. External genital organs of the female. 186

Waste disposal. Sewage. 215
Water. 53
　Water-borne diseases. 230
　Water purification 216
　Water supply. 214
White blood cells. Polymorphs and lymphocytes. 97
White matter. The nerve fibres which appear white because of a covering of fat in the myelin sheath. 137

Xanthoproteic test. A test for proteins using nitric acid. 50
Xerophthalmia. A condition in which the eye becomes clouded over because of vitamin A deficiency. 54
Xiphisternum. The cartilage extension of the breast bone. 28

Zygote. The organism resulting from the fusion of egg of sperm. 189